International Students in American Colleges and Universities

International Students in American Colleges and Universities

A History

Teresa Brawner Bevis and Christopher J. Lucas

INTERNATIONAL STUDENTS IN AMERICAN COLLEGES AND UNIVERSITIES
Copyright © Teresa Brawner Bevis and Christopher J. Lucas, 2007.
All rights reserved. No part of this book may be used or reproduced in any manner whatsoever without written permission except in the case of brief quotations embodied in critical articles or reviews.

First published in 2007 by
PALGRAVE MACMILLAN™
175 Fifth Avenue, New York, N.Y. 10010 and
Houndmills, Basingstoke, Hampshire, England RG21 6XS.
Companies and representatives throughout the world.

PALGRAVE MACMILLAN is the global academic imprint of the Palgrave Macmillan division of St. Martin's Press, LLC and of Palgrave Macmillan Ltd. Macmillan® is a registered trademark in the United States, United Kingdom and other countries. Palgrave is a registered trademark in the European Union and other countries.

ISBN-13: 978-0-230-60011-9
ISBN-10: 0-230-60011-5

Library of Congress Cataloging-in-Publication Data

Bevis, Teresa Brawner.
 International students in American colleges and universities: a history/Teresa Brawner Bevis, Christopher J. Lucas.
   p. cm.
 Includes bibliographical references and index.
 ISBN 0-230-60011-5
 1. Students, Foreign—United States—History. 2. Universities and colleges—United States. 3. Education, Higher—United States. I. Lucas, Christopher J. II. Title.
 LB2376.4.B48 2007
 378.1'9829—dc22
                                                                2007015268

A catalogue record for this book is available from the British Library.

Design by Macmillan India Ltd.

First edition: December 2007

10 9 8 7 6 5 4 3 2 1

Printed in the United States of America

Table of Contents

List of Tables	xi
Preface	xiii
Acknowledgments	xv
Introduction	**1**

1 Foreign Students and Centers of Higher Learning in Antiquity — 13

- Foreign Students in Ancient Greece — 14
- Roman Schools — 16
- Museums — 16
- Student Misconduct — 17
- Wandering Scholars in Medieval Europe — 19
- Foreign Students in the European Universities of the High Middle Ages — 20
- The Confessional Age — 26
- The Grand Tour — 28

2 Early American Colleges and Education Exchange — 31

- The Founding Fathers — 32
- Drawing Upon the German Ideal — 34
- From the European Perspective — 36
- Francisco de Miranda and the First Latin American Enrollments — 40
- Yung Wing and the Chinese Education Mission — 42
- The First Enrollments from Japan — 53
- The Beginnings of Immigration Policy — 55

3 The Early 1900s: Foreign Student Enrollments and Emerging Support — 59

The Foreign Student Census 1900–1930 — 60

In the Spirit of de Miranda: Growing Latin American Enrollments — 62

The Boxer Rebellion and the Return of the Chinese — 63
   A 1909 Foreign Student Services Survey — 65
   Tsing Hua College — 68
   Teachers College at Columbia University — 69
   "Slave Education" — 71

The Pensionados — 74

Enrollments from Other World Regions: France, Japan, and India — 78

Cosmopolitan Clubs — 81

The International Houses — 84

The Committee on Friendly Relations Among Foreign Students — 88

Female Foreign Students in the Early 1900s — 89

Foreign Students and Christianity — 92

The Shaping of Governmental and Institutional Policy — 93

The Institute of International Education — 95

Exclusions and Quotas: Immigration Policy Developments — 98

4 The World War II Years and Their Aftermath — 101

The Birth of the Fulbright Educational Exchange Program — 103

The Continuing Story of the Institute of International Education — 108

UNESCO — 111

Smaller Ventures — 112

	The 1948–1949 Foreign Student Census	113
	The National Association of Foreign Student Advisers	120
	Burden of Proof: Evaluating Credentials	123
	War and Postwar Immigration Policies	125
	Stranded in America	126
5	**Foreign Students, McCarthyism, and the Cold War**	**131**
	The McCarran Legislation	133
	Atoms-for-Peace: Deploying Knowledge	136
	The Bamboo Curtain	137
	Internationalizing America's Campuses	140
	Issues of Adjustment	142
	The 1955 Foreign Student Census	143
	The Training of Physicians	146
	New Organizations and Initiatives for Exchange	148
	The 1959–1960 Foreign Student Census	151
6	**The 1960s and 1970s**	**155**
	Innovations and Growth in Exchange Programs	159
	The IIE: New Contributions	161
	The South African Education Program	163
	The 1969–1970 Foreign Student Census	164
	Community Colleges and Foreign Students	167
	Emerging Research	171
	The 1979–1980 Foreign Student Census	172
7	**The Late Twentieth Century**	**175**
	Global Competition	175
	The Community College Alternative	177

	Making the Sale	182
	A Closer Look at International Students	184
	Attitudes and Characteristics of Foreign Students	187
	Reentry	188
	Intellectual Migration	189
	Enrollments in the Late Twentieth Century	191
	The 1999–2000 Foreign Student Census	192
	Growing Enrollments from India	194
	Borders and Taxes	195
8	**The Opening Years of the Twenty-First Century**	**201**
	International Students and the Rising Threat of Terrorism	201
	9/11: The Debate Continues	203
	Continuing Controversy Surrounding SEVIS	207
	Homeland Security and International Students: The Ongoing Debate	208
	Visa Restrictions	211
	The Enrollment Decline	212
	Responses to the Enrollment Decrease	216
	Enrollment Stabilization	219
	International Competition for Students	220
	The Bologna Process	222
	Looking Backward and Ahead	223
Epilogue: Challenges and Imperatives		**231**
	Security Questions in the Aftermath of September 11, 2001	231
	The "Crowd-out" Effect	235
	The International Student Enrollment Decline Reconsidered	237
	Reforming Immigration Law and the Visa System	238

Global Competition	240
The Need for Strategic Planning	241
International Students in America	242
Ambiguity in the Multiple Rationales for International Exchange	244
Looking Ahead	246
Notes	249
Bibliography	271
Index	281

List of Tables

4.1	Foreign Student Enrollment in United States Colleges and Universities 1942–1943	114
4.2	1948–1949 Foreign Student Enrollment: Top Ten Countries of Origin	118
4.3	1948–1949 Foreign Student Enrollment: Top Ten Institutions	118
5.1	1955–1956 Leading Nationality Groups in Foreign Student Population	144
5.2	1959–1960 Foreign Student Enrollments by Country of Origin	151
5.3	1959–1960 Institutions with More than 400 Foreign Students	152
6.1	1969–1970 Countries Represented by More than 1,000 Foreign Students	165
6.2	1969–1970 Foreign Student Enrollment: Top Ten Institutions	166
6.3	1969–1970 Foreign Student Enrollment: Top Ten States	166
6.4	Number of Foreign Students from OPEC Countries for Selected Years, 1954–1979	173
7.1	1999–2000 Leading Institutions by International Student Enrollment	193
7.2	1999–2000 Foreign Student Enrollments: Top Ten Places of Origin	193
8.1	International Student Enrollments	226
8.2	Countries of Origin of International Students in the United States, 2005–2006	226
8.3	Highest International Student Enrollments by Host States, 2006	226

8.4 Leading Host Universities for International Students in the United States, 2006 — 227

8.5 Leading Fields of Study of International Students in the United States, 2006 — 227

Preface

The broad topic of foreign students in American colleges and universities has generated remarkably few book-length works over the past three-quarters of a century, despite the fact that foreigners studying in the United States currently number more than half a million. They contribute—literally—billions of dollars to the American economy, and account for approximately 4 percent of all students enrolled in U.S. higher education.

Articles written about foreign collegians in American academe have tended to be relatively brief or highly specialized in nature. Most have appeared in limited circulation periodicals or "in-house" professional journals, thereby precluding chances of their attracting a large professional or popular audience. In the absence of more current alternatives, the few books on the topic constitute the most frequently quoted books in the field.

The earliest of these is a work edited and assembled by W. Reginald Wheeler, Henry H. King, and Alexander B. Davidson titled *The Foreign Student in America*, which was printed in 1925. The second book, published in 1955, was authored by Edward C. Cieslak and bore the title *The Foreign Student in the United States*. A third reference, *Foreign Students and Higher Education in the United States*, often depended on by present-day writers (though it first appeared in 1956 and is mostly outdated), was written by Cora Du Bois and published by the American Council on Education. Last is *The Foreign Student in the United States*, a monograph published in 1970 by Helen C. Clarke and Martha Ozawa.

The writing of this book was prompted by a desire to produce a more up-to-date, comprehensive historical overview of international student exchange. Such analysis certainly seems long overdue. In terms of what the present work contributes to the literature, however, a few caveats seem in order. First, the focus of this manuscript is exclusively on foreign students (undergraduate and graduate) sojourning in the United States. No attempt is made to address issues involving American students abroad, except as they relate to the origins of education exchange in the early years of the country's independence.

Second, while most of the narrative is historical in nature, the treatment is more "popular" than what would customarily satisfy professional historians. The target population or audience for this book, in any event, is intended to encompass faculty interested in global education and internationalism, student-services personnel who bear responsibility for helping international students on campus, academic advisors and planners, campus policy makers and analysts, and, of course, the general public (most especially those with firsthand experience interacting with international students).

Third, no attempt has been made to treat all aspect of foreign exchange in extensive detail. Topics dealt with lightly or deferred until later chapters include recent controversy over policies governing visas for foreign nationals, student recruitment initiatives in foreign countries, admission quotas, financial aid for noncitizens, "best practices" for designing and administering effective campus orientation programs for newly arrived foreign students, theories of student development, foreign student adjustment to the American cultural milieu, legal liability issues, promoting multiculturalism and intercultural awareness on campus, academic advising, English language training, and security issues. If these topics were explored in the detail they deserve, the discussion would necessitate more than a single volume.

Again, our intent is somewhat more circumscribed: to trace the history of foreign students in American higher learning. We leave it to others to interpret the full significance and meaning of the subject. For our purposes, it is sufficient to recount how and why hundreds of thousands of foreign students over the years have elected to come to the United States to pursue their studies, and to provide an overview of the resulting impact of their presence on American higher education.

Acknowledgments

A special note of appreciation is due our respective spouses, David and Kathleen, for their support during this project. We also thank colleagues, friends, and family members who provided research assistance, reviews, technical help, encouragement, and inspiration: Georgia Childers, Jared Smith, Lorraine Robert, Sydney Hayes Howington, Sherrie Starkey, Bart Cohen, Tom Bevis, Elizabeth Bevis, Louise King Brawner, and the late Thomas A. Brawner.

Introduction

"The pursuit of learning beyond the boundaries of one's own community, nation, or culture is as old as learning itself," observed Cora Du Bois, the author of a 1956 survey entitled *Foreign Students and Higher Education in the United States*. "It stems," Du Bois continued, "from the human capacity for curiosity and adventure . . . [and] reflects the ability of human beings to communicate with each other at varying levels and with varying sophistication across the barriers of social particularities."[1] Citing the many presumed benefits of international exchange and study abroad in a shrinking world, Oliver J. Caldwell, writing a decade or so later, affirmed, "We need a constant cross-fertilization of ideas between the many cultures of mankind. A logical function of education, in a world in which men and women everywhere have no alternative but to live as members of one community, is to serve as the medium whereby people become acquainted with each other."[2]

Closer acquaintanceship, according to a traditional view makes for better understanding and friendship across international borders and around the globe. Discussing the many advantages of hosting overseas students in America, one writer in 1925 avowed, "We can, if we will, send back each year to their many lands an army of ambassadors of good-will and helpful intercourse of international confidence, of hope and peace."[3]

If a foreign student has a fulfilling experience while studying in the United States, Helen Clarke of the University of Wisconsin declared in 1970, it "can contribute immeasurably to lasting friendship, respect, and the easing of world problems and tensions."[4] When a foreign country sends its future leaders to America for study, allegedly they are likely to form favorable impressions of the United States and so be more inclined in later years to support closer, more positive ties with their respective homelands. As anthropologist Ina Corinne Brown phrased it, "The direction taken by any particular country in the future may well depend on where its students go for an education and the kinds of experiences they have in the host country."[5]

But Brown added an important caveat. "While getting to know one another can, and often does, promote understanding and good will, it by no means insures such results. . . . Merely sending our students abroad or bringing students from other countries here are practices that may be considered useful preludes to understanding but are not in themselves any guarantee that international understanding or goodwill will be furthered." A positive experience is likely to prove highly beneficial. But a student's unsatisfactory experience is apt to have precisely the opposite effect.[6]

A participant at a 1948 conference on international student exchanges likewise sounded a cautionary note: "It is easy enough for men of good will to conjure up a glow of disembodied benevolence such as floats over all too many 'get-togethers' without coming to grips with the really very difficult business of human understanding. It is difficult enough when people belong to the same world and speak the same language . . . How much harder when they come from different lands and speak [diverse] tongues?"[7]

Other writers, agreeing, emphasized it would be an egregious oversimplification, in the words of two commentators, "to expect that a period of study in a foreign country will uniformly result in more favorable attitudes toward that country."[8] The truth of the matter, as many have pointed out, is that no one has yet discovered the extent to which study abroad actually fulfills the function of fostering amicable relationships between nations, whether it promotes mutual understanding, or contributes somehow to securing world peace. On the contrary, as many will add, if a foreign resident has a sufficiently unhappy experience, that student may return home disillusioned, embittered and utterly alienated from the people of the host country.

Anecdotal evidence describing both positive and negative experiences over successive decades is extensive. Typical in its effusive praise are the comments of a young girl from Finland who studied in the United States during the early 1950s. She found people to be helpful and friendly—despite occasional incidents where her hosts seemed brusque, even rude, and as she put it, "provincial." She thought she detected an "invisible curtain" between herself and her American friends that she could never push aside, a semi-transparent veil that suggested the expressions of friendliness shown her were somehow shallow and no more than superficial. Yet, on balance her experience turned out to be basically positive. "I had the pleasure to be invited often in American homes," she reported, "and I was impressed by the simplicity and warmth that characterized such meetings."⁹

Others—then and since—have not been so fortunate. Struggling against acute loneliness, an exchange student from India in the early 1920s lamented, "Oh, if I could only drop in occasionally at some friend's house and have a cup of tea, as we do in India, it would make all the difference in the world!"¹⁰ The same factors making for a difficult adjustment in the 1920s held sway decades later. A 1962–1963 study conducted at the University of Wisconsin, as a case in point, found international students discussing loneliness, social isolation, lack of friends, homesickness, difficulties in adjusting to Western food, unfamiliar routines and customs, hard-to-find housing—and a seeming lack of sensitivity on the part of citizens of the host country.¹¹ Americans, it was repeatedly observed, seemed "patronizing" when discussing international affairs. Yet, ironically, they appeared at one and the same time to be badly informed about the rest of the world and lacking accurate knowledge of other countries.

Americans, however, apparently were not the only host people to stand accused to provincialism. A Nigerian female teacher in England commented in 1970 on the incredible depth of ignorance allegedly manifest among the British. "I find many of the questions that people ask are intensely irritating," she confessed. "I have been asked if I knew how to use a knife and fork before I arrived in England, and if it is true that negroes have tails."¹²

Prior to the advent of the civil rights movement of the 1960s, America's largely unchallenged segregationist policies shocked and angered many foreign students, especially those of color who were most likely to be victimized by racial prejudice. "As for human relations," exclaimed one international visitor in 1952, "America is a failure . . . Here I am today, facing segregation. What a sham democracy! The pity of it is that students leave America disgusted. What will be their feelings toward America in their own countries where some day they may hold important positions in their government? Why can't America rid herself of hypocrisy and treat her citizens on an equal basis irrespective of color? This will win her the goodwill and confidence of other nations."¹³ A student from Uganda who had been refused food in restaurants fumed, "Mercilessly they send you away; and you go with your hunger and your anger, not knowing what to do." He asked rhetorically, "Is this the land of brotherly love?"¹⁴

Commenting on similar incidents, psychologist Otto Klineberg remarked, "The existence of prejudice and discrimination in many areas and among many people, and its unhappy consequences for interpersonal relations, and for the development of hostile attitudes toward the host country, represent a serious issue for students exchanges." He conceded, "Discrimination is indeed a reality which many students must face."[15]

Today popular opinion holds that Jim Crow legislation is little more than a relic of the past. Yet, few would deny that the underlying racism behind those hateful laws still flourishes to some extent in the darker recesses of American culture and social life. How much more time will be needed for the eradication of its vestiges remains to be seen.

Foreign students arriving on American shores to pursue their studies have always represented a great variety of nationalities, temperaments, and backgrounds. As Corinne Brown once observed, while the main purpose of international education has been education and this purpose is "worthy in and of itself," there is no escaping the fact that "whatever the avowed motive of the United States in encouraging students from other countries to come here, there have been certain expectations beyond the strictly academic ones."[16] Among them on the host country side must be included the desire to implant American-style democratic ideals in the hearts and minds of visitors, to promote human rights, and to celebrate the virtues of laissez-faire capitalism. But above all other motivating considerations, perhaps, has been an enduring desire to burnish America's image and to present an attractive face to the world. Generally speaking, the hope or expectation has been that through firsthand experience, international student visitors would come to appreciate both the people and the culture of the United States. Americans want to be liked. Hence, negative criticism or indictments of the country for its alleged racism, its social injustice, its imperialism, or any other shortcomings have invariably prompted a strong defensive reaction.

Throughout the twentieth century—albeit punctuated by intervals when nativism ran strong and the theme of a self-sufficient Fortress America protected by geography and benign neighbors held sway—the United States endorsed internationalism as an ideal (with varying degrees of enthusiasm) as it sought to build strong ties with its allies around the world. Anti-American sentiment ebbed and flowed at various junctures, depending more than anything else on the overseas popularity (or lack thereof) of specific U.S. policies and initiatives at a given point in time.

In the immediate postwar period, for example, world opinion toward America seemed mostly positive. But by the late 1950s, the symbol of the "Ugly American" had become a familiar icon, even in the midst of Cold

War tensions. U.S. military involvement in southwest Asia in the 1960s and '70s did little to win international support for America's initiatives either. A decade later, little had changed. A *Newsweek* poll conducted in 1983, for example, found America's global image once again severely tarnished. Most of the people surveyed worried about expanding U.S. global influence and feared that the nation's strong military presence actually had increased chances for war.[17]

At the midpoint of the twenty-first century's opening decade, anti-American sentiment in Europe, the Middle East, and Asia, which had surged again as a result of the U.S. war in Iraq, exhibited modest signs of abating, according to a sixteen-nation Pew Research Center poll conducted in 2005.[18] Nevertheless, the United States remained broadly disliked in most countries surveyed, especially throughout the Muslim world and in most of Western Europe. Only in Russia, Poland and India had relations with Americans registered improvement. The nation's image problem, most pundits agreed, had grown more serious than ever before in recent memory.[19] In the face of anti-Americanism abroad, renewed interest in exchange as one important means of improving relations with other countries might have been considered almost inevitable.

Historically, it is worth noting how the total enrollment of foreign students in the United States doubled in the first half of the twentieth century, rising from 3,673 in 1904 to almost 15,000 by 1946. Only four years later, the number of international students in America had doubled again, to 30,000. By 1951–1952, 30,462 college students from 126 different countries were enrolled in 1,354 colleges and universities—amounting to about one percent of the total collegiate enrollment for that school year. Between 1954–1955 and the year 2003–2004, the total of international students in the United States increased from 34,232, or 1.4 percent of total enrollment, to 572,509, representing more than 4.3 percent of all students registered in institutions of higher learning.[20]

As long as international student enrollments remained modest, both in terms of absolute numbers and as percentages of combined domestic and overseas enrollments, foreign exchange was apt to attract little official notice. It remained a decidedly peripheral phenomenon in American academe, little discussed, rarely remarked on. But by the mid-century mark, the presence on campus of ever-increasing numbers of students from countries around the world could no longer be perceived as a marginal enterprise. As more and more colleges and universities assumed responsibility for serving as host institutions for foreign scholars, they began to realize they needed to consider anew the multiple responsibilities involved.

In the early years of the twentieth century, at a time when many countries were signing voluminous cultural agreements for student exchanges, the U.S. government had mostly left the task of furthering cultural exchanges by default to private agencies.[21] An institutional response from the private sector was forthcoming for the first time in 1911 when the quaintly titled Committee on Friendly Relations Among Foreign Students was established in cooperation with the Young Men's and Young Women's Christian Associations. Its primary task was to render assistance to arriving foreign students at their points of entry into the United States, to help them complete immigration forms, clear customs, find transportation, and so on. Some few years later, in February 1919, the Institute of International Education (IIE) opened its doors as an information clearinghouse for international students. Finally, the National Association of Foreign Student Advisors (NAFSA) emerged out of a series of IIE meetings convened between 1942 and 1947. Article II of its By-Laws enunciated in stately prose what the fledgling organization aspired to accomplish:

> The purpose of this association shall be to promote the professional preparation, appointment, and service of foreign student advisors in colleges and universities, and in other agencies concerned with student interchange; to serve more effectively the interests and needs of exchange students; to coordinate plans for student interchange through comprehensive voluntary cooperation of all agencies and individuals concerned with exchange students; and in fulfillment of that purpose to initiate, promote, and execute such systematic studies, cooperative experiments, conferences, and such other similar enterprises as may be required to that end.[22]

Howard E. Wilson, writing from the perspective of the 1950s, felt that the arrival of great numbers of foreign students on campuses was "an outstanding phenomenon in the relationship of American universities to the contemporary American situation in world affairs. . . . " It was his feeling that institutions of higher learning had undertaken many types of international exchanges, but that more that a few schools had done so in an ad hoc, haphazard manner. "There is need for stocktaking," he judged, "for more thoughtful conservation of educational resources, for appraisal by each institution of what it can most successful and wisely continue to do in view of the rising importance of world affairs." He added, "Nowhere is this need more apparent than in the policies and practices by which we receive foreign students."[23]

Looking back a decade or so, one commentator in 1951 recalled that at a time when about 17,000 foreign students were attending classes in the United States, "the foreign student adviser on most campuses was a pretty weak person in terms of time allocation, budgetary provision, and time to do his job. Local success was more often the result of personal sacrifice and the use of off-duty time than of campus support and administrative cooperation."[24] By 1952 or thereabouts, an estimated 88 percent of all colleges had at least one foreign student advisor on staff.[25] Nevertheless, expanding numbers of college students from abroad had generated "a host of perplexing guidance and administrative problems," including the challenge of deciphering foreign academic transcripts in timely fashion, locating suitable student housing, designing effective campus orientation programs, solving health problems, supplying appropriate academic counseling, finding financial support when needed, and determining English proficiency, among other challenges.[26]

A certain urgency attached itself to resolving such problems. Whereas in the first quarter of the twentieth century college and university administrators had not shown much enthusiasm for encouraging students from abroad, by mid-century they had reversed their opinion dramatically once the realization dawned that foreign students had much to contribute to a college campus. By the end of the century when the number of foreign students exceeded half a million, or nearly 4 percent of the total collegiate enrollment, and hosting international exchanges had become a multibillion dollar enterprise, the stakes for failing to meet growing competition from other nations' institutions of higher learning were far too high to ignore.

The opening of the twenty-first century, not surprisingly, witnessed a redoubling of efforts to recruit foreign students to the United States. A 2001 handbook produced by NAFSA: Association of International Educators, for example, led off its appeal to potential student applicants by extolling the academic excellence of the American system of higher education: "The United States has one of the world's finest university systems, with outstanding programs in virtually all fields . . . [and] has something for everyone," the text declared. "U.S. universities pride themselves on being at the forefront of technology and educational techniques, and in making available to their students the best possible equipment and resources." The promotion closed with a reminder: "Experience in an international setting is a marketable commodity. Your long-term career prospects can be enhanced by an experience that develops self-confidence, independence, and cross-cultural skills—attributes that are in high demand with employers worldwide."[27]

Up until the terrorist attacks of September 2001, initiatives designed to encourage students from abroad to enroll in American colleges and universities were prosecuted vigorously. But in the aftermath of 9/11, the situation changed abruptly. Philip Altbach, a professor of higher education and director of the Center for International Higher Education at Boston College, spoke for many in characterizing the "inhospitable environment" that foreign students were now facing in America. "Coming to study in the United States," he observed, "has become an obstacle course, and prospective students abroad are increasingly leery of stringent, changing, arbitrary, and sometimes inconsistent government regulations regarding visas, reporting to government agencies, and the like." Students from developing countries, especially those from Middle East countries, he lamented, were being treated with disrespect by U.S. officials in their own countries. Worse yet, American university administrators responsible for international students were also reporting that a large number of students were being denied visas for no discernible reason whatsoever. Alternatively, they were delayed so long that the practical effect was to prevent students from coming to study at all.[28]

On May 18, 2005, the heads of forty-one national higher education associations—including the National Academy of Sciences, the American Council on Education, the Association of American Universities, and NAFSA: Association of International Educators—called on the White House and Congress to reduce visa headaches and hassles imposed on bona fide international students. Unless "the misperception that our country does not welcome these international visitors is dispelled," the leaders warned, "we risk irreparable damage to our competitive advantage in attracting international students . . . and ultimately to our nation's global leadership."[29]

Around the same period, a report from the Strategic Task Force on International Student Access, titled *In America's Interest: Welcoming International Students*, was released. Here too the nation's leaders were called on to streamline the visa process, to develop a national recruiting program to address the increased competition for international students by other nations, and to counter the first drop in applications to U.S. institutions of higher learning in more than a decade.

How well the nation's major policy makers will heed such warnings in future remains difficult to estimate at the present time. What is clear is that with 572,000 and more international students enrolled in the nation's institutions of higher learning, the United States is the largest host country on the globe, currently serving as temporary home to around 25 percent of all foreign students worldwide taken together. The U.S. continues to attract more international students than any of its competitors and, despite a recent drop in enrollment, more than the three largest ones combined. Interestingly, it used to be the case that a majority of international student visitors were European. Today, the majority come from developing or newly-industrialized countries such as China, India, Japan, Taiwan, and South Korea.

From almost any perspective, the importance of these foreign students on American campuses must not be underestimated. From a purely economic point of view, as previously noted, they infuse billions of dollars into the U.S. economy annually. Furthermore, an estimated two-thirds of these foreigners coming to pursue their studies (often with families in tow) reportedly pays all of their own expenses throughout the duration of their stay in the United States. Financially self-sufficient students such as these obviously represent an attractive and beneficial clientele for colleges and universities. It should come as no surprise, therefore, that their applications are much sought after nowadays.

Many advanced students—particularly in scientific and technological fields such as engineering, physics, and computer sciences—elect to remain in the United States after completing their academic studies, thereby filling key positions in the U.S. economy and enhancing the nation's global competitiveness. Even if they return home once their studies are completed, in the interim many have served as much-needed research and teaching assistants, or helped fill seats in specialized classes that might otherwise have been impossible to offer for lack of enrollees.

Large-scale political implications stemming from international exchanges are difficult to nail down with any precision. But socially and culturally, it is not difficult at all to believe that international students have an important leavening effect on U.S. collegians, given circumstances where social interaction of a meaningful sort actually takes place. By the very fact that they too are enrolled in courses of study, that they sit alongside their American peers in classes, that they share the same facilities, speaks to possibilities for learning and exchange. Their very presence, it may be observed, gives life and immediacy to the abstractions of "globalism" and "international-ism," of encounters with the Other. From closer proximity to one another can come appreciation for both fundamental differences and similarities, an appreciation capable of transcending national borders and separate histories.

Ultimately, if international students are not segregated off somewhere by themselves, both visitors and host students have much to gain from the exchanges that can and often do ensue. It is a cliché to speak of the world shrinking through advances in communication and transportation, of barriers to understanding and cooperation eroding away. But to the extent that the shrinkage is real and powerful, it seems not unrealistic to claim that now, during the early years of the twenty-first century, an education devoid of exposure to people from other cultures can hardly be considered a worthy education at all. The most effective corrective for narrow parochialism, it is frequently claimed by those who speak from personal experience, is exposure to people unlike ourselves. So, too, the best antidote to ethnocentricity is getting to know those whose backgrounds are instructively dissimilar from our own.

* * *

The narrative following opens up with a brief overview of how down through the centuries students have traveled to foreign lands to learn from famous teachers. Precedents for many present-day customs trace back to a time long before American higher education even existed. This opening chapter traces the influences that shaped student migrations from ancient times down to the advent of the modern world. Next to be considered are education exchanges in antebellum America, early European judgments (pro and con) on American colleges, and the growing willingness and desire of European students to come to American for their studies. Also noted are some of the colorful tales told by the first foreign students to study in America.

Subsequent topics treated in the chapters to follow include pioneering efforts by private organizations to provide support for foreign students, among them the first International Houses, the student-initiated Cosmopolitan Clubs, and the famous Committee for Friendly Relations Among Foreign Students. Also described are the beginnings of the Institute of International Education and various foreign student organizations. Broad developments throughout the span marked by two world wars and the Great Depression include the initiation of the Fulbright program, new legislation such as the McCarran Act, and the establishment of the National Association of Foreign Student Advisers.

From the century's midpoint on, major developments included shifting demographics, the emergence of different institutional imperatives, and an unprecedented influx of international students—for which most colleges and universities found themselves ill prepared—the whole taking place during a period of unprecedented social upheaval and challenges to existing systems, both within and outside the groves of academe. The aftermath of the 1960s and early '70s included new legislative, institutional and social changes profoundly affecting student migrations to the United States.

By the 1980s, colleges and universities were employing armies of practitioners to care for the increasing numbers of foreign students enrolling in American colleges. As the foreign student populations continued to expand, the professional field of foreign student advisement and support itself broadened in scope and influence on campuses across America. The decade of the 1980s also was noteworthy for the growing ability of community colleges to attract foreign students in ever-growing numbers. Subsequent years witnessed legislation and broad policy initiatives during the Clinton administration that greatly affected America's international student population.

Among the most important developments of the first few years of the twenty-first century were the attacks of September 11, 2001, on America. As popular fears for homeland safety mounted, demands for more stringent security measures were heard, most particularly restrictions on foreigners seeking permission to work or study in the United States. Also important in the early 2000s were major shifts in the composition of arriving foreign students in terms of ethnicity, preferred fields of study, countries of origin, and so on. How such changes will play out in years to come remains to be seen.

* * *

"Foreign student" refers to an individual from another country who is in the United States temporarily on a student visa, and who is registered at an accredited institution of higher education. For many years, the term was universally used and generally accepted. Later, like other phrases used by those involved with international affairs (such as "underdeveloped country" and "illegal alien") it was challenged by some who viewed the term "foreign" as possibly "politically incorrect." The term "international student" began to replace "foreign student" in collegiate circles. Many have argued, however, that the original descriptor was more precise. Organizations that have played fundamental roles in the development of education exchange—the Institute of International Education for example—continue to use the term "foreign students" when reporting census data or other information. The *Chronicle of Higher Education*, one of the leading resources for news and information in higher education, also gives preference to the term "foreign students."

Gary Althen, past president of NAFSA, wrote that "international" had little in its favor other than the blessing of those who believed "foreign" sounded unpleasant. Like many euphemisms, he contended, it was less precise than the term it was intended to replace. Rather than attempt to weigh the legitimacy of either term, this book uses both interchangeably.

1

Foreign Students and Centers of Higher Learning in Antiquity

Whenever a center of scholarship and learning arose in ancient times, the gathering of scholars it drew invariably included "foreigners"—that is, students not native to the immediate local area. Hence, in a very real sense some of the precedents for recruiting, accommodating, and supervising international students in today's institutions were established centuries ago. Many of the earliest issues pertaining to the hosting of nonlocals likewise still prevail in contemporary academe.

In ancient times, a journey of any appreciable length, whatever its purpose, was apt to involve considerable danger. Law and order were tenuous at best once the precincts of a town or city were left behind, and those who ventured into the vast forests or deserts did so at risk of life and limb. Attacks on itinerants by wild animals—wolves, wild boars, and bears in particular—were not unheard of in many parts of rural Europe as late as the seventeenth century. It also was essential to guard against bands of cutthroat brigands who hid themselves along a route awaiting an opportunity to prey on the unwary or unprotected. Unpaved roads and footpaths might unexpectedly turn impassable in bad weather, leaving hapless travelers stranded in the wilderness. Unreliable bridges posed the same problem. Inns and other types of shelter where one might find food and lodging along the way often were far and few between.

Sea voyages carried their own travel risks, not the least of which was the possibility of capsizing in one of the sudden storms for which the Mediterranean, as a case in point, has always been notorious. Thus, embarking on a long journey by land or sea was not something to be undertaken casually, particularly by inexperienced adolescents traveling to a foreign university.

Typically, people who did brave the hazards of travel banded together whenever possible for both protection and companionship en route. (One thinks immediately of Chaucer's *Canterbury Tales* with its depiction of a group of wayfarers making their way on a pilgrimage.) Fortunately, despite the many hazards and hindrances, most travelers (religious mendicants, merchants, artisans, barristers, adventurers, soldiers of fortune, and so on) managed to reach their intended destinations without mishap. Among them were would-be students who had chosen to leave their familiar home surroundings to pursue a course of a study—or some small portion thereof for the less serious-minded—in a distant locale. Foreign or international students in this sense have been a significant presence in higher learning since its beginning, the rigors of travel notwithstanding.

Foreign Students in Ancient Greece

While earlier civilizations undoubtedly hosted foreign students, the ancient Greeks were among the first to capitalize on drawing them in from faraway regions. One of the distinguishing features of classical and Hellenistic Greek education, in fact, was the mobility enjoyed by all parties concerned. Most instruction was conducted without benefit of an actual physical facility, which meant "classes" could convene wherever convenient and could be moved from place to place as circumstances dictated. A small area in a city's open marketplace or *agora*, or perhaps within the shade of the colonnades of a nearby temple usually sufficed for a teacher's classroom needs. Whether situated in a building or convened in the open, these so-called "rhetorical schools" proved to be extremely durable, historically speaking, numbering in the hundreds well into the late Hellenistic era. The best-known among them drew students from far beyond the host city, lending an unmistakable "international" flavor to an otherwise "local" learning community.

Among the earliest testimony to the participation of "foreign" students in Greek higher learning is a comment from Socrates, noting (in the fifth-century BCE) the arrival in Athens of the Sophists (*sophisai*, or "teachers of wisdom"). He observed that the teachers were accompanied by numerous protégés—youths from distant regions who traveled with their masters as they made their way from city to city seeking new students. "Most of Protagoras' followers," Socrates observed, "seem to be foreigners; for these the Sophist brings with him from the various cities . . . charming them . . . with his voice, and they, charmed, follow where the voice leads."[1]

> In the Hellenistic world, Athens in particular began to attract students from various countries to its higher education institutions. As Athens became increasingly a university town, "Foreign potentates . . . vied with one another in endowing her with beautiful buildings, while students of all ages and nationalities thronged her streets and drew inspiration from her associations."
>
> J.W.H. Waldon, 1909

Plato, Socrates' famous student, may have instituted one of the first instances in the ancient Greek world of a physical "school" devoted expressly to higher learning. The Platonic Academy, its name deriving from the hero Academos, was founded around 387 BCE in Athens. It is recorded that Hippias built a circuit wall to enclose the area to be used for schooling; while the Athenian Cimon planted the grounds with trees to provide beauty and shade. In later years, the complex was furnished with an elaborate circuit that enclosed the lecture area. Also, groves of trees were planted to supply beauty and shade.[2] The Academy evolved, over an extended period, into a major classical center of learning, arguably the foremost of its kind. It reportedly attracted foreign students from throughout the Mediterranean world.

Three other great philosophical centers, each of which was similar to the Academy and attracted students from faraway locations, were founded to Athens. These were Aristotle's *Lykeion*, or "Lyceum" established in 335; the "Porch" or *Poecile* of the Stoics, founded in 310; and the *Kepos*, or "Garden" founded by the Epicureans in the year 306. Many lesser scholarly centers were created in subsequent years, in sufficient numbers to further mark Athens as the premier center of intellectual learning and activity. It was a reputation that the city enjoyed through the waning years of the Hellenistic era.

As early as the second century BCE, the attendance rolls of the city-state's schools of philosophy and rhetoric attest to the admission of foreigners in comparatively large numbers.[3] Among the more than a hundred or so foreigners matriculated in any given year, only two or three Roman names typically were recorded. The names of attendees from the great cities of Asia Minor and the Aegean islands, however, appeared on the rolls frequently.[4]

Roman Schools

In Roman times, it was customary for youths with a scholastic bent to first attend one of the private primary or elementary schools common throughout the imperium. The grammarians and rhetors, who could be found in practically any place of cultural importance, offered adolescents more advanced instruction. Finally, aspiring scholars could elect to attend a rhetor's school. These institutions proved long-lived, existing well into the fifth century BCE in the East, at Cos, Ephesus, Smyrna, Pergamon, Antioch, Constantinople, and Athens. The last was an especially attractive center, both for wayfarers en route to other destinations and to youths whose primary intent was to enroll in a school kept by some famous teacher. Among them was the great orator and rhetorician Cicero, who reportedly spent six months in Athens studying under Demetrius the Syrian. Countless others, before and after, submitted themselves to the tutelage of some scholar of great repute.

In the West, schools of outstanding distinction flourished later at Limoges, Toulouse, Marseilles, Arles, Lyon, Treves, Milan, Naples, and of course, preeminently, the Eternal City of Rome.[5] In virtually all instances, a school's reputation for scholarly distinction was sufficient to attract students from outlying regions, wherever Roman rule prevailed and sometimes even beyond imperial borders. Study abroad had become a well-established practice for young Romans of wealth, ambition, and adventuresome spirit.[6]

Museums

In addition to philosophical and rhetorical schools, another type of educational institution that flourished in Greco-Roman times was the *Mouseion*, or Museum ("Dwelling of the Muses"), which served as a scholarly research center for higher learning. The most famous among these was the great Museum at Alexandria, founded in 280 BCE or thereabouts under the royal patronage of Ptolemy Soter I. In its prime, the institution attracted some of the most eminent scholars of the day—poets, men of letters, geometers, astronomers, historians, physicians, dramatists, critics, and grammarians.

Museum pensioners and temporary residents alike lived in community near Ptolemy's palace. They paid no taxes and apparently had no particular duties to perform. Most of them resided in luxurious private apartments at royal expense but took their meals together in a single vast dining hall. They reportedly spent most of their time engaged in scholarly pursuits, after the fashion of professorial retirees and scholars on sabbatical. "In this populous Egypt of ours," wrote Timon in his *Satirical Poems*, poking fun at their pretensions, "there is a kind of bird-cage called the Museum where they fatten up any amount of pen pushers and readers of musty tomes who are never tired of squabbling with each other."[7]

Some accounts report that the Museum boasted several magnificent botanical and zoological gardens as well as a world-famous library, one of at least three within the host city. Together with its annex, the *Serapeum*, the Museum is said to have contained no less than 120,000 volumes, according to the dialogue drawn up by its third librarian, Callimachus, between 260 and 240 BCE. The Alexandrian library was the largest of its kind in the world. Details about its actual whereabouts, layout, holdings, organizations, administration, and physical structure, however, are matters of conjecture.[8] About all that can be said with any confidence is that the Alexandrian edifice (and other rival institutions resembling it at Rhodes, Antioch, Ephesus, Smyrna, and Pergamon) was intended to sustain a community of scholars, more or less imitating the spirit of the Athenian school of the early Pythagoreans and likewise at a later date the Academy, the Lyceum, and the Garden of the Epicureans.

The Museum was basically a center for scholarly inquiry and research, not for advanced instruction. Its scholars-in-residence usually sought seclusion in order to pursue their studies; and they were under no obligation to deliver lectures or otherwise offer instruction. Nevertheless, many did teach. Perhaps it was inevitable that distinguished scholars would attract disciples and finally consent to instruct students who sought academic mentors, as was the case of Dionysius of Alexandria, who studied under the grammarian Aristarchus, or Apion, who became the pupil of Apollonius. In any case, by the fourth century BCE Alexandria was a great university city, particularly famous for the study of medicine and the hearing arts. The Museum, perhaps in concert with two or three other "universities," was attracting students from far and wide—from as far away as Cappadocia, for example, as well as from other distant locales.[9]

Student Misconduct

Evidence suggests that when large numbers of young, out-of-town students gathered, in Alexandria or any other major center of learning, the atmosphere was one of rowdiness and frivolity, a state of affairs brought about by hordes of unruly youths with too little discipline and far too much free time on their hands. Many students were away from home and parental jurisdiction for the first time, and insufficient supervision could and did lead to trouble with local authorities for any number of infractions: public intoxication, brawling, petty theft, ribald practical jokes, blatant sexual immorality, extravagant betting on horse races, and excessive gambling among them.[10] Many well-behaved students undoubtedly resided in any given metropolis (St. Gregory Nazianzen and St. Basil in fourth-century CE Athens and Zacharias Scholasticus and Severus of Antioch in fifth-century CE Beirut serve as examples of exemplary student piety and decorum), far better remembered are the gangs of disreputable foreign students whose lawless antics vexed official guardians of law and order.

Municipal and imperial authorities tried without great success to rein in the most egregious offenders. In the year 370, for example, the Emperor Valens issued an order requiring foreign students arriving in Rome or Constantinople to register their academic credentials and testimonials to good character with the local tax assessor before seeking admission for study in a school. The requirements set forth clearly prefigure modern-day admissions and immigration procedures for foreign students:

> Whosoever comes to Rome for the purpose of study must first present to the head of the Board of Censors a letter from the judges of his province (from whom he in the first instance received permission to come), containing the name of his city, and a statement of his age and qualification. As soon as he arrives he must signify to what studies he intends to devote himself. The Board of Censors must be kept informed of his residence, in order that they may see that he follows the course he has laid out for himself. . . . Those who devote themselves assiduously to their studies may remain in the city until their twentieth year; after that time, whoever neglects to return to his home of his own accord shall be made to do so by the city prefect, under disgrace.[11]

Notwithstanding, judging from the frequency with which similar imperial edicts and injunctions were issued, continuing throughout the reign of Theodosius in the fifth century CE, it appears the emperor's proclamations were not nearly as effective as beleaguered authorities might have wished.

Wandering Scholars in Medieval Europe

A succession of distinctive European civilizations followed the final eclipse of the Western Roman Empire in the sixth century CE, beginning with the Merovingian and Carolingian cultural revivals of the seventh and eight centuries, respectively, and culminating at the apex of the Middle Ages in the twelfth and thirteen centuries. Medieval students, like their predecessors centuries previously, were highly mobile in their quest for education. As the historian Charles Haskins notes, "Men moved in leisurely fashion from place to place in search of eminent masters, careless of curriculum or fixed periods of study or degrees. . . ."[12] Students were, as he phrases it, "singularly mobile and singularly international . . . Bologna has its English archdeacons and German civilians, Paris its clerks from Sweden and Hungary, as well as from England, Germany, and Italy. Even the cathedral schools drew from beyond the Alps and across the narrow seas. . . ."[13]

Itinerant students were not always regarded favorably, particularly when they were foreign. Most were young, nearly penniless, and more often than not did not necessarily speak the language or local dialects of the countries through which they passed. Poorly dressed and usually dependent on others for food and provisions, wandering scholars were looked down upon as "blightsome" and "squandering" youths "born for toil and sadness . . . ragged and continuously begging."[14]

One of the most enduring images of the student sojourner, and certainly one of the most widespread in early modern England, reflected the prevailing medieval belief that student vagabonds were habitual liars. An eighth-century CE critic of wandering students complained that they roamed from monastery to monastery, depleting both their host's table and patience with their tale-telling.

> Behold him now come from the Italian frontier, and a good fresh tale all about pilgrimage or captivity, entering the house with humbly bowed head, and lying hard til all the poor host's poverty goes into the pot and on to the table; that host will be a well-picked bone in a day or two.[15]

Judgments about the *vagantes*—wandering scholars—remained substantially unchanged four centuries later. Observes J. A. Symonds, itinerant clerks of the 1100s—students, ex-students, teaching masters even—moved from town to town in great numbers in search of learning and, still more, of adventure.[16] They were nominally unbeneficed clerks who for the most part led very un-clerical lives: "Far from their homes," Symonds observes, "without responsibilities, light of purse and light of heart, careless and pleasure-seeking, they ran a free and disreputable course."[17] Another account by a twelfth-century monk commented, "They are wont to roam about the world and visit all its cities, till much learning makes them mad; in Paris they seek liberal arts, in Orleans, classics, at Salerno, medicine, at Toledo, magic—but nowhere manners and morals."[18]

So-called "Goliardic" poetry (possibly an obscure allusion to Goliath the Philistine) penned by wandering clerks illustrates their characteristic preoccupation with the attractions of wine, women, and song. "In the public house to die is my resolution," runs one verse. "At life's dissolution that will make the angels cry, with glad elocution, grant this toper, God on high, grace and absolution!"

Declarations of extreme poverty frame another typical theme: "These torn clothes that cover me, are too thin and rotten," the narrator mourns. "Oft I have suffered cold, by the warmth forgotten." The poet then importunes, "Take a mind unto thee now, like unto St. Martin; cloth the pilgrim's nakedness, wish him well at parting." But abject poverty by itself was insufficient to induce homeless students to embrace a more settled lifestyle. "We in our wandering," the poet affirms, "eat to satiety, drink to propriety ... laugh till our sides we split, rags on our hides we fit; jesting eternally, quaffing infernally...."[19] This sort of iconoclastic, rebellious student lifestyle was endlessly celebrated in Goliardic-like poetry for centuries and was standard fare by the time of the great universities at the height of the Middle Ages.

Foreign Students in the European Universities of the High Middle Ages

The medieval university as an institution of higher learning originated as an outgrowth of advanced courses of instruction offered within municipal cathedral church schools, beginning sometime in the tenth and eleventh centuries. During the formative period of the university, almost any academic gathering where students and masters were drawn together in pursuit of learning was referred to as a *studium* ("place of study") or less frequently, a *discipulorum*, an association of persons devoted to scholarly pursuits. When it happened that a teaching master became sufficiently renowned to attract students from beyond the immediate area of the cathedral, however, the term that gradually came into usage was *studium generale*. It served, basically, to distinguish a larger or more "general" place of study from a strictly local institution. (Only much later did the term *universitas*—originally referring to any sort of formal association or corporation—come into more restrictive usage to refer to a teaching-learning community.) Hence, in a fundamental sense, the first great universities of medieval Europe owed their existence to the growing numbers of foreign students who flocked to city cathedral schools for instruction. A local school was just that—a strictly local affair. But a school whose reputation or location had grown to the point where it was sufficient to attract students from great distances now was deemed a true "university."[20]

Between 1100 and 1200, European culture was affected by an influx of new knowledge—learning conveyed for the most part through the hands of Arabic scholars in Moorish Spain. The recovered texts of Aristotle, Euclid, Ptolemy, and other prominent classical writers (works long lost to European scholars) had an electrifying effect on medieval European culture. Essentially, the new knowledge now engulfing the Western world was breaking down the narrow bonds of the old cathedral and monastery schools. Something new was in process of emergence, and it would ultimately evolve into the university as its own institutional type.

Among the dozens of urban church schools that dotted the map or Europe by the 1100s, several had already begun to acquire reputations for particular areas of academic specialization. The cathedral church school at Toledo, for instance, was best known for its translations of Greek and Arabic scientific treatises. Medicine was taught at Salerno and Montpelier. At Bologna and Orleans, canon and civil law were the main specialties. Chartres was known for its outstanding school of liberal arts.

Teachers of classics and philosophy at the latter institutions, such as the brothers Bernard and Thierry, began attracting students from distant locales, youths drawn by the prospect of studying with leading lights of the day. As early as 991 a monk of Rheims by the name of Richter described in some detail his experiences as a foreign enrollee at Chartres, where he studied the *Aphorisms* of Hippocrates of Cos under the cathedral school's resident teaching masters.[21] A century later, Peter Abelard (1079–1142), the great scholastic philosopher and theologian, was helping to establish Paris as Europe's premier center for theological studies. He quickly gathered an avid following of students, many of whom had come to France from other states.

Only in Germany were schools of higher learning slow to develop. Hence, in the twelfth and thirteenth centuries, German students who lacked opportunities to study at home were obliged to seek learning abroad, usually in France or Italy. (Circumstances changed in later years, however.) Indicative of official German policy and support for education at the time was a provision contained in the *Authentica habita* of 1158 in which Emperor Frederick Barbarossa guaranteed scholars studying in Italy safe conduct in imperial lands.[22]

No two *studia generalia* evolved in exactly the same way. The most widespread pattern, particularly throughout northern and central Europe, was for the teaching masters affiliated with a cathedral's *studium* to organize themselves into distinct guilds or "nations" (*nationes*)—very much resembling those that dominated any other trade or profession in medieval society (for example, vintners, cobblers, brewers, glass-blowers, and so on). Each nation reflected a shared ethnic or regional identity of its members and a common vernacular language.

Toward the middle of the thirteenth century, there were four officially acknowledged nations at Paris: the French, Norman, English, and Picardian. Curiously, the French nation enrolled professors coming from both northern France and the Low Countries. The English nation accepted members from Holland, the German states, Denmark, Sweden, Norway, and the Hungarian and Slavic regions.

Contemporary accounts suggest that sectional and ethnic rivalries often deeply divided a school's teaching masters and students. From the outset, frictions within and among the nations were pronounced. Feuds and factional rivalries were especially well represented in Paris. One chronicler reported:

> They wrangled and disputed not merely about the various sects or about some discussions, but the difference between the countries also caused dissensions, hatreds and virulent animosities among them, and they impudently uttered all kinds of affronts and insults against one another. They affirmed that the English were drunkards and had tails; the sons of France proud, effeminate, and carefully adorned like women. They said that the Germans were furious and obscene at their feasts; the Normans, vain and boastful; the Poitevins, traitors and always adventurers. The Burgundians they considered vulgar and stupid. The Bretons were reputed to be fickle and changeable.... The Lombards were called avaricious, vicious, and cowardly; the Romans, seditious, turbulent, and slanderous; the Sicilians, tyrannical and cruel; the inhabitants of Brabant, men of blood, incendiaries and brigands and ravishers; the Flemish, fickle, prodigal, gluttonous, yielding as butter, and slothful.[23]

Some university nations were comprised by teaching masters, as at Paris. In fledgling universities elsewhere, however, the nations that began to make their appearance in the early 1200s did so as student associations, or *collegia*, that is, fraternal organizations of "assemblies" of foreign students banded together for collective security and protection.[24] Far from home and undefended, out-of-town students who shared a common vernacular language or home region invariably found it expedient to band together for mutual protection (from other students but also sometimes from hostile local authorities).[25] At Bologna, for example, the four "Italian" nations drew students from Rome, Campania, Tuscany, and Lombardy. Apprentice-students coming from abroad (from "over the mountains") of course affiliated with the appropriate non-Italian alternative. Guilds or nations for foreigners included the French, Spanish, Provencal, English, Picard, Burgundian, Poitevin, Tourainian, Norman, Catalonian, Hungarian, Polish, German, and Gascon. Once again, the important role played by international students and student nations serves to illustrate how critical so-called "foreigners" were to the existence and character of the evolving medieval university.

The average age of an entering student was no more than fifteen or sixteen in most cases. The first challenge awaiting him on arrival in the city was to find suitable lodging. Selecting a place to live often posed a daunting challenge for any youth without relatives or compatriots to offer counsel, especially as the university offered those admitted little more than a place to attend lectures. Practically no guidance was given regarding how to get settled into the school's routines. The odds of being cheated or otherwise taken advantage of were therefore great, especially for youngsters not yet in full possession of their legal rights and untutored in the local language and customs. Unsurprisingly, as a result, unwary foreign students were frequently overcharged for miserable accommodations by opportunistic locals.

Some European universities, however, did make provision for helping indigent students. Among the earliest was the so-called "College of the Eighteen" in Paris. In return for a small allowance and a bed at the Hospice of the Blessed Mary, students were asked to offer prayers and take turns carrying the cross and holy water before the bodies of deceased patients. The hospice restricted its hospitality to students of exemplary behavior; and every student's comportment was evaluated weekly. Residency was limited to a year, with the possibility of a year's renewal at the discretion of local authorities. By this time most foreign students would have become acclimated to the town and were capable of picking their own lodging. Over time, the college's initiatives in providing shelters for young boys gave rise to much more elaborate endowed residences for university students.[26]

Likewise challenging was fending off the clamoring hordes of students that hung about the inns or taverns awaiting each new arrival to the city. Hired by masters as "chasers" or recruiters, each member of these groups would set upon a newcomer, urging that he enroll under a particular teacher, all the while deriding the merits of that master's competitors.

Once a room had been hired and a teacher selected, the new student's next step was to present himself before the representative of the nation appropriate to his place of origin, where he would swear an oath of allegiance to the ranking officials of the university and to the members of his nation. Once his matriculation fee was paid, the student was admitted forthwith as an apprentice-scholar in good standing.

In its formative stages, the medieval university was hard pressed to enforce rules and regulations intended to maintain control over its own student body. Statutes of 1442 at Heidelberg proclaimed: "No student shall presume to visit the public . . . gaming houses. . . . No one shall presume by day or night to engage in gaming or to sit or tarry by night or otherwise for any time in a brothel or house of prostitution." Students were further cautioned against dancing during Lent or going about in "unseemly" fashion wearing masks, engaging in jousts, swearing in public, or uttering blasphemy. A decree at the same institution in 1466 warned vigorously against "clamor or insolence" and made it illegal for anyone to compel a first-year student to "throw filth" at other students or masters.[27]

The *Manuale Scholarum* of 1481 at Heidelberg includes a rather horrific account of the many indignities heaped on newly admitted students by their peers. Rites of passage went well beyond ridicule or mere name-calling. In several instances it is recorded that an initiate was smeared with excrement, soaked in urine, or forced to ingest a strong cathartic, all the while being whipped or beaten about the head. University officials tried repeatedly to abolish the more brutal forms of hazing, but to little avail. Typical was a statute issued by Leipzig in 1495: "Each and every one attached to this university is forbidden to offend [first-year students] with insult, torment, harass, drench with water or urine, besprinkle or defile with dust or any filth, mock by whistling, accost with a terrifying voice, or dare to molest in any way whatsoever. . . ."[28] More often than not, students simply ignored all such official proclamations and threats.

Reminiscent of the unruliness that prevailed a millennium before among students in the rhetorical schools of the Hellenistic period, medieval university students—a vocal percentage of them, at least—were a rowdy bunch. As medievalist John Baldwin observes, "Animated by youthful exuberance, protected from the local police, at liberty to pursue their studies anywhere, these young scholars often lived at the margin between the urban underworld and decent society. By the twelfth century students who wandered from town to town in search of famous masters formed an *ordo vagorum*, an order of wanderers, in impious parody of the monastic and priestly orders of the day."[29]

Indicative too of a certain lack of seriousness on the part of some students is a letter written by an irritated father to his errant son. "I have recently discovered that you live dissolutely and slothfully, preferring license to restraint and play to work and strumming a guitar while the others are at their studies, whence it happens that you have read but one volume of law while your more industrious companions have read several." He concludes: "Wherefore I have decided to exhort you herewith to repent utterly of your dissolute and careless ways, that you may no longer by called a waster and your shame may be turned to good repute."[30]

Similar in tone is a missive written by a parent who had been informed by his son's teacher that the young student might benefit by a parental admonition to mend his ways. Not wishing to disclose the source of his information, the parent wrote, "I have learned . . . from a certain trustworthy source . . . that you do not study in your room or behave in the school as a good student should, but play and wander about, disobedient to your master and indulging . . . in certain other dishonorable practices which I do not care to explain by letter."[31] The customary exhortation to the youth to apply himself more diligently follows.

Clearly, universities were untidy affairs, almost always turbulent, contentious, and marked by strife between contending factions forever engaged in disputations over this issue or that. The presence of so many students and their masters within certain European communities must have seemed to the local townsfolk a mixed blessing at best. Yet it cannot be denied that a congregation of so many young men, drawn from such disparate backgrounds, carried with it multiple benefits—cultural, economic, and even political. At the very least, the presence of so many young masters from divergent backgrounds and cultures lent a decidedly cosmopolitan flavor to the local environment.

The Confessional Age

Writing in the year 1513, Italian political theorist Niccoló Machiavelli registered an eerily prescient judgment. "Whoever examines the principles upon which that religion Christianity is founded," he declared, "and sees how widely different from those principles its present practice and application are, will judge that her ruin or chastisement is near at hand." Just four years later Martin Luther posted his famous ninety-five theses on the church door at Wittenberg. Within a single generation nearly half of Europe had abandoned Roman Catholicism for Protestantism, and the medieval unity of European Christendom had been irreparably shattered.[32]

Confessionalism, both Protestant and Catholic, had a profoundly deleterious impact on Europe's institutions of advanced learning. Internecine struggles born of the Reformation that led to bloody Huguenot wars in France, the Thirty Years' War in Germany, and relentless sectarian strife in England basically ushered in a long period of educational somnolence and stagnation. Given the circumstances that prevailed across Europe between the late sixteenth century and the end of the eighteenth, it seems remarkable that universities continued to function at all.[33] Impoverished by war, and lacking the energy and resources needed for major reform, most lower schools could do little to extend educational work of any kind. As for the universities themselves, they had all they could do simply to survive. Many did not. As an Oxford vice-chancellor, surveying the devastation brought about by constant civil unrest and warfare, lamented in 1643, "We will hang our harps on the willows and now at length bid a long farewell to learning."[34] Not surprisingly, the span of time extending from the early 1500s to the late 1600s was not at all conducive to international exchanges or the hosting by universities of foreign students from distant lands. Apart from the obvious hazards for anyone trying to make his way across Europe's war-torn landscape, relentless conflict had transformed most universities from sanctuaries of independent thought into instruments for buttressing the doctrinal orthodoxy of particular sects or denominations (consistent with the principle that a university was obliged to pledge its allegiance to whatever faith the governing ruler embraced). Likewise, throughout the seventeenth century, all who attended a given school were obliged to profess the officially sanctioned faith that institution espoused. Thus, only Catholic stalwarts were permitted at institutions of learning loyal to the Mother Church; all taint of heresy would be ruthlessly rooted out and suppressed.

Precisely the same considerations dictated that only Calvinists could attend a Calvinist school. Lutherans for their part would not venture to attend anywhere but at a Lutheran university, and so on. In an age of rigid intolerance where theological differences were considered matters—literally—of life and death, some easy transition from one school to another, as had been common in an earlier time, was now a virtual impossibility.

The English, however, seem to have earned a predisposition for finding educational opportunities in foreign lands under any conditions. Despite the hardships and dangers of traversing the Channel and traveling on the Continent, hundreds of English men and women journeyed abroad in medieval Europe each year. This vigorous tradition prompted one chronicler to comment on the English people's love of travel and the number of them abroad: "the peple of that londe is dispersede a brode thro alle the worlde, trawenge alle the worlde to be a cuntre to theyme."[35]

Apart from diplomatic, merchant and military traffic, many medieval Englishmen traveled abroad as scholars. Pilgrimage provided many with an excuse to indulge their curiosity about the world beyond England, and large numbers of scholars traveled to universities on the Continent. Although travel for educational purposes was a well-established practice, these scholars were not the direct predecessors of early modern educational travelers. By the late fifteenth and early sixteenth centuries, the changing emphasis of education stimulated by the civic humanism of the Renaissance spawned a new type of educational traveler, who visited other countries not only for scholarly accomplishment but also to fit himself for service to state and prince.

Education was the great leveler in English society; and a good education often became the key to success for a low-born citizen. Travel was one way ambitious students could supplement what formal education they had. Numbers of "nongentry" educational tourists began to grow in the second half of the sixteenth century. For those without bloodlines, money, or influence, the experience of travel abroad could serve as an apprenticeship for a respectable position in state service. The experience and education gained during a tour through Europe often became the cornerstone of a young man's later career in the service of the "publike weale."[36] In his quest for patronage and employment a young man emphasized any experience or education that he had gained through a lengthy tour abroad. The practice, in fact, became almost a prerequisite for inclusion into English society.

The Grand Tour

Most treatments of international exchange limit themselves to accounts of students making their way to distant locales to enroll in some celebrated institution of higher learning. Mentioned less often is an interesting variation on the theme afforded by the so-called "Grand Tour"—a distinctly English invention that entailed educational student travel but not actual enrollment anywhere for formal study.[37]

The term Grand Tour was introduced by Richard Lessels in his 1670 book *Voyage to Italy*. Additional guidebooks, tour guides, and the tourist industry were developed and grew to meet the needs of the twenty-something male and female travelers and their tutors across the European continent. No less a personage than Queen Elizabeth I (1533–1603) judged that a grand tour of Europe's cultural centers might well provide a fitting culmination of an adolescent's liberal education.[38] It would serve, she claimed, as a useful rite of passage, a means of testing the ingenuity and initiative of a young man undergoing the rigors of travel abroad, prior to his posting to some appropriate position. As Francis Bacon (1561–1626) explained the monarch's view, her injunction was to send abroad "some young men of whom good hopes were conceived . . . to be trained up and made fit for public employments. . . ."[39] Bacon hastened to add that the young men who embarked on a European tour "traveled but as private gentlemen" without any royal stipend or other support. How they comported themselves on their travels, it was thought, would serve to reveal their character and hence suitability for public employment. "As by their industry their desserts did appear," Bacon remarked, "so were they further employed and rewarded."[40]

> The grand tour was designed for young men who had already completed their formal education at home but who wished to smooth off their rough edges and acquire a veneer of cosmopolitanism.
> Edward Charnwood Cieslak, *The Foreign Student in American Colleges*, 1955

The typical grand tour itinerary did not vary much between the sixteenth and early nineteenth centuries. Always included were extended sojourns to France, Switzerland, Italy, Germany, and the Low Countries. A visitor on tour, with help from a manservant, was expected to have prepared himself to appreciate what each region had to offer. "He must have such a servant or tutor as knoweth the country," Bacon advised. "Let him carry with him also some . . . book describing the country where he traveleth, which will be a good key to his inquiry."[41]

The typical eighteenth-century traveler made his way by horse coach, often traveling at night in order to avoid the cost of renting a room at a tavern or way station en route. Also commonplace were extended "courtesy visits" at British consular offices located in most of Europe's larger cities. The presence of so many young visitors at one time often proved a major burden on their hosts. In 1772, for example, the envoy in Florence, Frances Colman, complained, "I have hardly had one hour to myself this week by reason of concourse of English gentlemen that are here at present."[42] A tourist would not carry much money due to the risk of highway robbers, so letters of credit from their London banks were presented at the major cities. Due to the travel expenditures English tourists contributed a great deal of money to businesses abroad, causing some English politicians in the interest of the local economy to be very much against the institution of the Grand Tour.

For many of the young men sent off to see the world, the freedom suddenly thrust upon them was positively intoxicating. The prevailing view of parents, however, tended to be one of acceptance of misbehavior elsewhere that would never be tolerated at home. By the same token, the great majority of youths who went abroad were serious-minded individuals anxious to benefit as much as possible from their travels and exposure to foreign cultures.

Having experienced the Grand Tour several years previously, one commentator reflected favorably on his experiences abroad: "From both trouble and pleasure I assuredly derived instruction and on many occasions I have essentially profited from this my first entrance into the world." He continued, "To many perhaps the experiment might have been dangerous; to me fortunately it proved of the most real and lasting advantage. The variety of scenes through which I had lately passed, the society into which I had been introduced, and the manners and information which I had acquired, made me on my return extremely acceptable to all my old friends, and procured me the acquaintance of many, to whom I otherwise had small pretension to be known. . . . "[43] Interest in the Grand Tour waned toward the end of the eighteenth century. The precedent for including foreign travel as an integral part of higher learning, however, was by now well established.

2

Early American Colleges and Education Exchange

Given the prestige of Europe's academic institutions, it is hardly surprising that America's colonial and antebellum colleges failed to attract many foreign students. Not until the mid-1800s did students from places such as India and China begin to enroll in sizable numbers in the United States. In common with the rest of the world, American students usually looked to the academic centers of England or Germany when seeking the most prestigious higher learning available. The practice of studying abroad during the early years of America's independence thus became the precursor to all education exchange to follow and helped set the conditions that in time created a two-way migration.

During America's early years, the idea of education exchange was not universally embraced by either the public or politicians. Sending students to Europe for higher education was an unwelcome acknowledgment that American institutions did not yet hold the same prestige; and for many Americans, any reinforcement of ties to Europe was unfavorable. One fitting illustration of popular opinion dates to 1750, when the Common Council of Philadelphia made a donation to Benjamin Franklin's Academy. The Council optimistically expressed its hope that with the opening of Franklin's new school, no young person would be "under necessity for going abroad" for their education.[1]

In fact, Franklin himself was a strong proponent of European higher education and maintained a number of ties with several European academic centers. He had spent considerable time at Göttingen, Germany, for example, observing German higher education and intending to return home with progressive ideas that could help make American institutions more competitive and attractive.[2]

The Founding Fathers

Benjamin Franklin's affection for European higher education was not shared by all of the founding fathers, however, at least with regard to sending young Americans to Europe's universities. Perhaps surprisingly, given his reputation as a "Renaissance man," Thomas Jefferson was one American who adamantly opposed Americans studying in Germany, or for that matter, any other country except their own. In a letter from Paris to J. B. Bannister Jr. dated October 15, 1785, the future president conceded that medical students might need to go to Europe, but insisted that all other American youth could pursue their studies at home without much sacrifice of quality. Jefferson strongly supported the building of a national identity distinct from European tradition and ideology. He observed that "an American, coming to Europe for education, loses in his knowledge, in his morals, in his health, and in his habits." He continued, "The consequences of foreign education are alarming to me, as an American."[3] And in another letter to his nephew Peter Carr, Jefferson remarked: "There is no place where your pursuit of knowledge will be so little obstructed by foreign objects, as in your own country, nor any, wherein the virtues of the heart will be less exposed to the weakened."[4]

Jefferson did concede great respect for European academics nevertheless. Writing to Governor Wilson C. Nicholas of Virginia, on November 22, 1794, he described the universities of Edinburgh and Geneva as "the two eyes of Europe,"[5] and he even attempted to persuade the Virginia legislature to transplant the University of Geneva to America as a state-funded institution. Further, Jefferson wrote to President George Washington about the transplant idea in 1795, declaring that, should the University of Geneva be brought to the United States, it would satisfy the needs of most Americans for advanced learning. Washington decided to take a more nationalistic view. Relocating the University of Geneva to America might prove to be harmful rather than beneficial in his opinion, for by so doing the distinctly European habits and principles, good and bad, would certainly be brought along with it.

Washington's views were further illuminated in his letter to Governor Robert Brooke of Virginia of March 16, 1795, in which he stated his belief that there was a danger in sending American youth abroad among other political systems when they had not well learned the value of their own. "It is with indescribable regret that I have seen the youth of the United States migrating to foreign countries in order to acquire the higher branches of erudition and to obtain a knowledge of the sciences."[6]

The Revolutionary War, nearly two decades past, had been hard on American colleges, and higher education was in a period of recovery. The war had not only damaged buildings, but also had resulted in fewer enrollments, smaller endowments, and weakened reputations. Perhaps the biggest impact of the war, however, was on the purpose and focus of American education itself:

> The war may have begun as an effort to redefine the economic relationships between mother country and colony. It may have become a movement for independence, but before it was over it was also a movement for democracy: a full-bodied statement to the effect that in America man counted for more, took less account of his superiors—indeed frequently denied their existence achieved whatever distinction his own ability and the bounty of the land allowed him, looked any man in the eye and knew him as an equal before the law and before God. . . . The spirit of this rising democratic tide overtook the colleges. . . .[7]

Americans were quickly developing a distinctive national identity, as Jefferson had hoped, and were supporting the growth of the young nation's higher education programs as never before. Between 1782 and 1802 nineteen colleges were chartered, more than twice as many as had been founded during the previous 150 years.[8] In light of America's newfound interest in its colleges and universities, President Washington urged educators nationwide to develop improved systems of all phases of higher education, and encouraged legislators to support those changes.

Washington also had a dream of establishing a national university. The idea had first been championed by Benjamin Rush, a prominent physician and patriot of Philadelphia.[9] The idea gained its strongest endorsement from George Washington, and in his first and last annual messages to Congress in 1790 and 1796, the president solicited the legislators' support for the establishment of a University of the United States. Washington argued that a national university that would serve to shape America's patriotic citizens and able civil servants, enabling the United States to develop a class of men free from provincialism and sectionalism. It would save young Americans from having to go abroad, promote their attachment to republican forms of government, and provide the basis for a first-class university, competitive with those of Europe. While Washington's idea received some support, his dream of a national university was never realized. Nevertheless, his enthusiasm for the improvement of higher learning in general helped to encourage policies that would result in the rapid growth and advancement of America's colleges and universities over the next several decades.

Perhaps it is ironic that while the actions of Jefferson and Washington were initially designed to thwart study abroad and keep students home, the resulting improvements to higher learning in the United States ended up attracting students and scholars from everywhere else. The path to education exchange with Europe, and the rest of the world, soon became a two-way road.

Drawing Upon the German Ideal

Plans proposed by America's founding fathers to advance education in the new republic typically arose from their common desire to emulate the institutions of Germany. German universities were considered the most advanced in the world throughout the eighteenth and nineteenth centuries. A significant component of their popularity was that they exhibited a marked degree of elasticity in their curriculum, offering freedom from the restraints and lack of choice usually encountered by students in other systems. This sense of creativity and new opportunity that academic flexibility promised drew scholars from all over Europe and Asia, as well as from America, making Germany a center of international higher learning. Poet William Longfellow, for example, preferred Göttingen over Oxford, which had not yet adopted a welcoming stance toward American students.[10]

The first academic degree awarded to an American student in a German university was that of M.D., conferred upon Benjamin Smith Barton by the University of Göttingen in 1799. Between 1791 and 1850, about a hundred Americans had enrolled in German universities. Of particular note among those students was George Ticknor, who studied in Göttingen and went on to indelibly mark American higher education, although the changes he initiated would not be fully institutionalized until well after the Civil War.

By its own admission, Harvard College before the Civil War was a provincial academy. It did a good job of preparing students to become respectable citizens, but by comparison with European institutions it was, in the opinion of many who had studied abroad, indifferent to the advances in knowledge that were beginning to reshape the world.

As a young man, Ticknor had made the decision to study at a European university. He had been intrigued by Madame de Staël's celebrated book *De l'Allemagne*, which described the intellectual excitement surrounding the renaissance then taking place in German philosophy and literature. Also intrigued by a pamphlet that described in detail the University of Göttingen's revolutionary system of study, Ticknor determined to make Germany his destination.

Twenty months at Göttingen convinced Ticknor that America was considerably behind Germany in its ability to provide higher learning. He was alarmed to discover, for instance, that his Greek tutor at the university, who was only two years older than he, was much further ahead of him in scope and precision of knowledge. In a letter to his father Ticknor wrote, "What a mortifying distance there is between a European and an American scholar. We do not yet know what a Greek scholar is; we do not even know the process by which a man is to be made one."[11]

After completing his studies, Ticknor returned to the United States to become a member of Harvard's faculty. Once there, he resolved to try to incorporate German ideals into the American system. He crafted a plan for a full-scale departmental program on the German model, including lectures in the prescribed languages, and presented it to Harvard's president, John Thornton Kirkland. When he found that the prevailing conditions stood in his way, he set about promoting an across-the-board reform of the entire academic program. "A great and thorough change must take place in its discipline and instruction" to make sure, he added—with an asperity not likely to endear him to colleagues—that the college would at least "fulfill the purposes of a respectable high school."[12] President Kirkland acknowledged the institution's need for reform and was receptive to Ticknor's plan. But the faculty at large simply was not. "The resident teachers," Ticknor told a friend, "declared themselves against all but very trifling changes." Ticknor's plan to reform Harvard began to materialize only under President Charles William Eliot (who filled the office from 1869 to 1909) a half-century later, transforming it into a university in the broadest sense of the word. Colleges and universities across the nation followed Harvard's example, and George Ticknor would be later referenced as "the originator of the university idea in America."

From the European Perspective

Washington's visionary mandates, to finally bring American higher education to a respectable level and convince the world's academic community that the United States could compete academically with Europe, were slow to develop. George Ticknor's seedling proposals likewise required time to take hold. In the meantime, assessments of American higher education, especially in the first half of the nineteenth century, were often less than glowing. Thomas Hamilton, a Scottish novelist, for example, chronicled the story of a tour of American colleges in 1831 in a publication titled "Education and National Characteristics." Here he describes Harvard College:

> The buildings, though not extensive, are commodious; and the library—the largest in the United States—contains about 30,000 volumes; no very imposing aggregate. The academical course is completed in four years, at the termination of which the candidates for the degree of Bachelor of Arts are admitted to that honor, after passing the ordeal of examination. In three years more, the degree of Master may—as in the English Universities—be taken as matter of course. . . . The number of students is somewhat under two hundred and fifty. These have the option of either living *more academico* in the college, or of boarding in houses in the neighbourhood. No religious tenets are taught; but the reigning spirit is unquestionably Unitarian. In extent, in opulence, and in number of students, the establishment is not equal even to the smallest of our Scottish Universities. . . .[13]

Hamilton also broached the issue of racial prejudice in America and provided an anecdotal account of the tribulations of one foreign student of African descent to illustrate his point. Hamilton's account described the son of a Haitian general, a youth who had come to America for instruction and enlightenment, having long heard tales of the opulence and energy of New York City. A mulatto, the young man was reported to have a pleasing and refined manner, superior intelligence, and a reasonably good education. As the privileged son of a general, he had been accustomed to receiving all of the deference due his rank at home in Haiti. In his experience, racial prejudice had never been an issue.

On his arrival in New York, weary from travel, he inquired for the best hotel, and immediately made his way there, but was refused admittance. Similar attempts at several other hotels met with the same result. At length, the Haitian student was forced to take up residence in a miserable lodging-house kept by a local American woman of African descent. The pride of the young Haitian "who, sooth to say, was something of a dandy, and made imposing display of gold chains and brooches" had been wounded by the ordeal, and the experience confirmed his conviction that, in this country, he was regarded as a degraded being, with whom "the meanest white man would hold it disgraceful to associate."[14] The young Haitian's dreams of an American education and an enlightening experience in New York were gone, according to Hamilton's account, and the student returned to his native island by the first conveyance, never to return.

Hamilton's conclusion from this and other stories he encountered was that American colleges were not yet prepared to educate foreigners. In Hamilton's view, the young Haitian, or for that matter any student of color, would be better off choosing a European school, instead. In England, Hamilton claimed, the student would have felt quite secure, and "if he have money in his pocket, he will offer himself at no hotel, from Land's End to John O'Groat's house, where he will not meet with a very cordial reception." He stated further that churches, theaters, operas, concerts, coaches, chariots, cabs, wagons, steamboats, railway carriages, and air balloons would be at the student's disposal. "He may repose on cushions of down or of air, he may charm his ear with music, and his palate with luxuries of all sorts. He may travel *en prince,* or *en roturier,* precisely as his fancy dictates, and may enjoy even the honors of a crowned head, if he will only pay like one."[15]

Isaac Fidler took an even dimmer view of American higher education. Fidler, an English clergyman, classical scholar, and linguist, emigrated to America in 1831 with his family and a servant to seek greater teaching opportunities. His 1832 article "An Immigrant's Anecdotal View of the State of Learning in America" records his disappointment with his reception. In the course of trying to secure a teaching position, he discovered that even Harvard College had no appointment for a "Sanscrit [sic] or Persian scholar." He encountered the same lack of opportunity everywhere, and some American educators even suggested he take up a new profession. Embittered by his failure to find an appropriate post, and having toured most of the prominent institutions of the day, Fidler grew increasingly disillusioned with United States higher education in general. He declared that, in a country where higher learning was not yet sufficiently valued and where "expulsion from a public institution entails no disgrace, nor disqualifies for any kind of business or pursuit, it appeared to me improbable that much attention to instruction could be secured. I therefore asked if such a system of education could lead to eminent acquirements?"[16]

He further asserted that Americans were rude, and accused them of being strangers to refinement and of taking pleasure in humbling and scorning those whose manners differed from their own. "If your children mix with [Americans]," Fidler said, "their conduct towards yourself will be so contaminated with Republican principles as will be a source of hourly vexation."[17]

While early criticisms might have had merit, over the next decades American colleges and universities emerged as credible centers of learning and instruction. New institutions were built, old ones improved, and support from legislative and civic leaders was substantial. It was a period of adopting and adapting European (especially German) precedents for research and flexibility, in tandem with solidifying America's own unique character, energy, and spirited innovation. In the latter part of the nineteenth century, Europe and the rest of the world began to take notice.

An 1876 article by Ferdinand Eduard Buisson, titled "French Views of American Schools," illustrated how American higher learning was improving in the eyes of a critical world. Buisson, a French educator and politician, and later co-winner of the Noble Peace Prize, served as a professor of pedagogy at the Sorbonne and as Director of Elementary Education in France. He was also well-known as the editor (1882–1893) of the *Dictionaire de pedagogie et d'instruction primaire.*

Buisson and his assistants had been appointed by the French government to examine and report on American education. While the main focus of their study was public schooling, Buisson's comments reflected a new European appreciation of American education in general. "If any people ever used this 'power of education,' or united its destinies to the development of its schools," Buisson reported, "or made public instruction supreme guarantee of its liberties, the condition of its prosperity, the safeguard of its institutions, that is most assuredly the people of the United States."[18] According to Buisson, what in the beginning might have seemed a momentary burst of enthusiasm had assumed the force of a profound and settled conviction. No longer the work of philanthropists or of religious societies alone, American education had become a public charge for which the citizenry willingly paid a tax higher than that of any country in the world with which it could be fairly compared.

Emil Hausknecht, a German educator and professor at the University of Japan in Tokyo and later the director of the second *Realschule* in Berlin, offered similar words of praise for American colleges in his 1893 article, "German Criticism on American Education." Hausknecht crossed the United States visiting places of interest and institutions of higher learning, keeping notes on his observations. He reported having found an open and exciting expression of ideas on the part of leading persons in science and promoters of popular education:

> He who has seen with his own eyes . . . will know that in that country, though it is partly still in primitive development, yet everywhere progressing with gigantic strides, and disregarding Old World prejudices, something higher rules than "filthy lucre," which, under the present circumstances of civilization, can scarcely be called "filthy" any longer.[19]

Hausknecht's report further described the American college as a kind of intermediate institution between Germany's *gymnasium* and university, representing the upper grades of the former and the first two years at the latter. In terms of facilities, he described the typical American institutions as having dormitories, lying remote from the noisy din of cities, surrounded by park-like grounds. "In the colleges physical exercises are very popular. While, with few exceptions, no beer drinking or smoking is found, yet the student's life here is very gay, full of fun and pranks,"[20] asserted Hausknecht. He further reported that unlike German universities, most American colleges served the purpose of transmitting knowledge. The chief object of German universities—to guide students to independent scientific production—was entirely outside of the scope of the American college in his view.

While Hausknecht's comments acknowledged that there was still room for improvement, he recognized, as did most of the world's educators by then, that a passion for learning had seized the American population. By 1900 the status of American education was in the ascendancy, and United States institutions were beginning to be viewed as attractive destinations for foreign students seeking not only knowledge, but also opportunities for research and freedom of personal expression.

Michael E. Sadler, an English professor who worked with both Trinity and Oxford universities, recorded this comment in "Impressions of American Education" just a few years later than Hausknecht, in 1902:

> At rare intervals in the history of a nation there comes a great outburst of physical and intellectual energy which, with over-mastering power, carried forward the masses of the people, together with its leaders, in an

exhilarating rush of common effort. In the United States of America such a movement is in progress today....[21]

The atmosphere in the United States was one of equality and of independence. According to Sadler, American higher education had profoundly changed the organization of universities through the development of what were called "elective studies"—a direct outgrowth of America's appreciation of individuality.[22] Praise for American colleges continued to spread among educators and the public at large. By the turn of the twentieth century, students from all over the world began to look to the United States as an increasingly attractive venue for higher learning.

Francisco de Miranda and the First Latin American Enrollments

Despite the negative early reports made by Europe's leading educators about American universities, at least a handful of foreign students actually arrived well before America's colleges came of age. Admissions rosters of American antebellum colleges attest to the presence of some foreign students in attendance toward the close of the eighteenth century, and a slightly larger number in the opening years of the nineteenth. At a time when United States institutions still had no formal means of serving the unique needs of foreign students, and when travel could still be extremely challenging, the first nonnational enrollees were clearly adventurers. Many of them left indelible marks on American education. Often colorful, some of their stories have reached legendary status in the history of foreign student exchange.

The Venezuelan revolutionist Francisco de Miranda was a historic figure in the South American struggle for independence from Spain and may have been the first foreign student to enroll in an American college. Clearly, he was a pioneer of sorts. By nature Miranda was an adventurer. As an officer in the Spanish army he served under Bernardo de Gálvez in the Spanish attack on Pensacola (1781), when Spain was an ally of the rebels in the American Revolution. From this experience was born his fascination with America and its educational opportunities was born. He visited Philadelphia and Boston, where he met George Washington, Alexander Hamilton, and other notables. He also traveled widely in Europe, particularly in Russia, where he became a favorite of Catherine the Great. He later fought in the French Revolution.[23] His desire for education, however, brought him back to the United States. In 1784 he became Yale University's first foreign student.

Miranda returned to Venezuela and soon took a commanding position with the patriot forces. He became dictator for a short time, but after increasing misfortunes, including the loss of Puerto Cabello to the famous liberator Simon Bolívar and a destructive earthquake in Caracas, he surrendered (in 1812) to the Spanish. Unfortunately for Miranda, Bolívar and other patriots, angered by his capitulation, seized him and turned him over to the Spanish authorities who, failing to honor the terms of surrender, deported him to Cádiz, where he was confined in a dungeon for the remainder of his life.[24]

Also among the first foreign student "pioneers" to arrive from Latin America was Venezuelan Fernando Bolivar of Caracas, nephew of Simon Bolivar. Born in 1810 in Caracas, Fernando Bolivar came to the United States at age twelve to attend Germantown Academy in Germantown, Pennsylvania, while his famous uncle was fighting to create independent nations in South America. When Fernando's father was killed in the fighting, Simon Bolivar, who had no children of his own, adopted his young nephew. After five years in preparatory school, Fernando considered going to West Point. But because of his admiration for Thomas Jefferson, he decided instead to attend the newly founded University of Virginia. His admission marked the arrival of the university's first Latin American student.

Simon Bolivar sent a letter to the university faculty detailing how he wanted Fernando to be educated. A typed copy of the letter, along with other material relating to Bolivar's days at the school, today reposes in Alderman Library's manuscript collection. Among other things, Simon wanted Fernando's education to include modern languages, "not neglecting his own," he explained. He expressed his appreciation for the university's attention to Fernando's particular needs, with a notation that the professors were "very eminent men secured by Jefferson for the express purpose of teaching at his university."[25]

Sadly, before the year was out, Fernando Bolivar was forced to leave the university when the commercial house that had been handling his funds went bankrupt. James Monroe generously offered to let the distressed student live in the brick cottage on the university grounds where Monroe had his law office, but Fernando decided he should return to his native Venezuela. Once there, he first served as personal secretary and confidante to his uncle, then went on to build his own distinguished diplomatic career.[26] Bolivar was important not only because he was the first Latin American student at the University of

Virginia, but also because his attendance marked the beginning of a long-standing academic link between the university and Latin America.

Another of America's first foreign students was Mario Garcia Menocal, the son of a Cuban sugar plantation owner. He studied initially at the Institute of Chappaqua, New York, and then the Maryland College of Agriculture. He later attended Cornell University, where he graduated in 1888 with an engineering degree.

Armed with youthful ideals, an ardent appreciation for Western education, and visions of a career as a political leader, Menocal returned to Latin America after his graduation, but not yet to Cuba. He first associated himself with his uncle, Aniceto Menocal, on the commission for the study and construction of the Nicaragua canal route. Later, he served with the military, earning distinction in the battles of Yerba de Guinea, La Piedra, and La Aguada, as well as in the capture of the Fort Loma de Hierro. Menocal eventually became president of Cuba. His first term was characterized by a constructive policy that included achieving closer relations with the United States, an ideal clearly bolstered by his American college experience.[27]

Although America's first foreign college students were Latin American, students from China and other parts of Asia soon followed, and in much greater numbers. Asia soon became, and has remained, the largest exporter of students to the United States.

Yung Wing and the Chinese Education Mission

Largely untouched by the advent of modernity elsewhere, nineteenth-century China featured abject poverty, social unrest, and a corrupt imperial administration. While the country's population continued to increase, its productivity had not kept pace. China's per capita cultivated land and wages had been declining since the twelfth century—and this decline gained momentum after the seventeenth century.[28] Defeated by Great Britain in the First Opium War (1839–1842) and by Britain and France in the Second Opium War (1857–1860), by the nineteenth century, China's resources were seriously depleted. It was increasingly apparent that the imperial government had to take dramatic steps to bolster the economy, a challenge requiring the gathering of knowledge and technology from the West. A few extraordinary individuals played important roles in pioneering the first programs of education exchange between China and the United States. One was Yung Wing, Yale class of 1854.

Yung Wing (a.k.a. Rong Hong, Jung Hung) was the first Chinese ever to be awarded a degree from an American institution of higher education. He was born in Nan Ping, a province of Guangdong, in 1828, where fate also had placed a small Christian school run by Mrs. Gutzlaff, the large and robust wife of an English missionary. Yung Wing's parents enrolled him for classes there when he was just seven years old. Many years later, in his autobiography, Yung Wing described his first meeting with Mrs. Gutzlaff and recounted having " . . . trembled all over with fear of her imposing proportions."[29] He did not like the little school, but it was not long before he progressed to the Morrison School, another American missionary school in nearby Macao, this one run by Reverend Samuel Robbins Brown, a graduate of Yale. Yung Wing described Dr. Brown as an outstanding teacher, " . . . cool in nature, versatile in the adaptation of means to ends, gentlemanly and agreeable and somewhat optimistic."[30]

One day in August 1846, Reverend Brown made an announcement to his class: Because of health reasons, he would soon return to America and wanted to take a few students with him, so that they could complete their education there. He asked the students who was interested in accompanying him. In what turned out to be a life-changing decision, Yung Wing stood up. He and two of his classmates, Wong Shing and Wong Foon, were selected to accompany Dr. Brown.

The boys were excited at the prospect of going to America. Dr. Brown's stories and teachings had inspired their dreams of education—and stirred mental images of an exciting adventure crossing the Pacific. In one of Yung Wing's student essays, titled "An Imaginary Voyage to New York and Up the Hudson," he constructed a colorful tale of sailors and sea storms. Soon, he and his classmates would actually make that voyage, accompanied by Dr. Brown and bound for Monson Academy in Massachusetts.

In 1847, Samuel Brown and his young protégés departed China during a fierce storm, on a ship called *The Huntress*.[31] They arrived at the academy ninety-eight days later. Wong Shing soon returned to China because of illness, and Wong Foon left to study in Scotland, leaving only Yung Wing to attend Yale. There, the Reverend Charles Hammond, also a Yale graduate, befriended the ambitious youth. With guidance from Drs. Brown and Hammond, and armed with a scholarship generously provided by the Ladies Association of Savannah, Georgia (Hammond's sister was a prominent resident of Savannah), Yung Wing became the first Chinese ever to enroll in an American institution of higher education.

In 1851, Yale's freshman class consisted of ninety-eight Americans, and of that group, there were few students of color, no women—and Yung Wing. He was regarded as something of a curiosity by his classmates, as he kept his topknot and wore a traditional Chinese tunic his entire freshman year. Yale legend tells of Yung Wing being a loner who had little social interaction, although he was a common sight around campus as he worked regularly at both the dining hall and the library to help fund his tuition. He was an excellent student, but nearly failed differential and integral calculus, which, according to his autobiography, he "abhorred and detested."[32] He compensated for his mathematical shortcomings by winning the Yale Prize for English Composition, much to the chagrin of many of his native English-speaking classmates.

Yung Wing often took long walks and held lengthy discussions with classmate Carrol Cutler, who would later become president of Western Reserve College (Case Western Reserve) in Cleveland. Sharing a mutual interest in ensuring that the rising generation in China would have opportunities for Western education, Yung and Cutler came up with the idea for an education mission. They drafted a plan for systematically sending small groups of young Chinese to study in America over a period of several years, and hoped someday to be able to present the idea to the Chinese government.

It took time for their plan to be heard. Yung graduated from Yale in 1854 (two years after obtaining his American citizenship) and soon after left on a 151-day trip home, to reunite with his family and begin his career. His return to China, however, was accompanied by startling realization. When the ship docked in Hong Kong, a Chinese-speaking pilot came aboard to give the passengers instructions, and to Yung Wing's shock, he had trouble understanding him. He discovered that his native language skills had dramatically slipped from lack of use—a revelation that later led to the addition of Chinese classical studies to the mission's curriculum. Yung had to study language for six months before he could take the Confucian examinations required for entering the ranks of officialdom.

Not only had Yung Wing lost language skills, but because he had left China as a boy, he had few friends left and felt somewhat alienated from Chinese culture. Rather than trying to fit back into traditional society, however, he persisted in engaging himself in various projects and activities to promote China's modernization, an effort that naturally served to further alienate him from most of his still-conservative colleagues.

Yung Wing returned to America—which was now in the throes of a civil war. Genuine feelings of patriotism for his adopted country moved Yung to try to help. Finding his way to the local recruitment office, he volunteered as a soldier for the Union army, but to his deep regret was declined. Unmoved by his claims of citizenship status, army officials regarded him as Chinese and "not really American." Returning again to China, Yung entered into the service of the Qing (Ch'ing) dynasty, where his English skills could be put to good use.[33]

It was now 1868, and the Burlingame Treaty had just been signed by the United States and China, setting the stage for China to choose the United States, bypassing England, as the country to which she would send students for Western studies. Article VII of the document affirmed: "Chinese subjects shall enjoy all the privileges of the public educational institutions under the control of the government of the United States."[34] The treaty was named for Anson Burlingame (U.S. minister to China from 1861–1867). After leaving his ministerial post, Burlingame became China's ambassador-at-large and lobbied in the West to promote educational exchange between China and the United States. With his help, the Burlingame Treaty was drawn up and signed by the two countries. With the momentum generated by the treaty, Yung Wing could now set about implementing the plans he and Carrol Cutler had so long ago devised.

Despite opposition from Confucian conservatives who favored China's continued isolation, Yung Wing found sympathetic ears in the Qing dynasty, among them the Viceroy Li Hongzhang, who wanted to see China modernize after humiliating defeats dealt by European powers. Li, along with Viceroy Zeng Guofan, who was also in favor of opening channels to education exchange with the West, memorialized the signing of the treaty to Emperor Tongzhi:

> Article VII of the new peace treaty with America states that from now on Chinese who wish to study in [American] government-controlled schools and colleges will be treated the same as citizens from the most favored nation.... For these reasons, your ministers are seeking approval to establish a bureau in Shanghai, to recruit bright young boys from coastal provinces, at the rate of thirty per year, and a total of 120 in four years.[35]

When Yung's proposal was accepted, he wrote that he was "treading on clouds and walking on air."[36]

In 1872 Yung Wing returned to America to set up the project. According to the plan, students would travel to America in four equal contingents, with one group to embark each year. Some of those selected for the journey were as young as ten years old (the average age was twelve and a half); these youngsters were sent early to receive preparatory training before entering college. The Chinese government allotted each student a total of fifteen years to complete his education. If the first and second installments of students proved to be successful, everyone agreed that the plan would be continued indefinitely. Chinese teachers were to travel with the students, to maintain the students' knowledge of Chinese language and traditions while they were in the United States.

The Imperial Court appointed Chen Lan-Pin, a conservative Confucian with no knowledge of English, to be the first Commissioner of the Chinese Education Commission in Hartford, Connecticut, and Yung Wing was named Deputy Commissioner. Together, they constituted the first permanent mission China ever sent abroad.

As Deputy Commissioner, Yung Wing was responsible for selecting the first group of students to come to America. At the time, the journey from China to the United States was long and dangerous, and the requirement that students be away from home for a period of years made it a challenge to find talented and willing applicants. In fact, before the boys could be signed up however, the parents of each were required to sign an agreement to accept the will of destiny should the child become ill or die during his journey or years of study. Most of the students were recruited along the coastal areas, and twenty-five of the first detachment of thirty boys were from Guangdong.

The initial detachment of Chinese students left Shanghai on August 11, 1872.[37] Their ship docked in San Francisco on September 12, and ten days later the students and their sponsors had crossed the country to arrive in New England. The group was led by Commissioner Chen and included two instructors of Chinese classics and one English interpreter.

Housing was provided through the generosity of local New England families. Birdsey Northrop, secretary of the Connecticut State Board of Education, called for citizens to host the boys in groups of two and four, and 122 eager families in Connecticut and Massachusetts volunteered for the task.[38] The project was, in general, warmly received. The years following the Civil War were a time of tolerance toward minorities and immigrants, and Hartford was an intellectual center enjoying postwar expansion and prosperity.

The students settled in and before long were excelling in their new environment. Their academic achievements were matched by their victories on the baseball diamond and in the ballroom, as their great "southpaw" pitcher, Liang Tun-Yen, led the "Orientals" to many victories. (Liang later served as the last Minister of Foreign Affairs in the Qing Dynasty).

For the two consecutive years (1880 and 1881) that Yale defeated Harvard in crew (rowing) races on the Thames River in Groton, Connecticut, Chinese mission student Chung Mun-Yew served as coxswain for the Yale varsity crew. Several stories have been told about his crew exploits, including one published in the *Hartford Courant* in 1912:

> Famous in Yale annals as the coxswain of the Yale shell which distanced Harvard in the race of 1880, Chung was a favorite among his classmates. He was a bright student who never lost his temper and who was never known to swear, except on one occasion. That was during the race with Harvard in 1880. Toward the finish, the little coxswain broke out with "Damn it boys, pull!" The boys did and Yale won the race.[39]

The mission students were popular in social circles and, by all accounts, quickly adjusted to American culture. A classmate at Hartford Public High School, William Lyon Phelps, recalled that at dances and receptions "the fairest and most sought-out belles invariably gave the swains of the Orient preference." Elsie Yung, Yung Wing's granddaughter, said in an interview many decades later that when she was a girl in Shanghai, her father's ex-Mission friends would tell her, "I used to dance with Mark Twain's daughters."[40]

The Chinese students' stay in America could not have happened at a more exciting time in the nation's history, as it coincided with a great period of scientific and technological innovation. The Chinese students witnessed the invention of Alexander Graham Bell's first telephone (1876), and Thomas Edison's phonograph (1878) and incandescent lamp (1879). They also attended the Centennial Exhibition in Philadelphia, where samples of their homework on display in the Educational Pavilion won merit awards from the Board of Jury. The students' many accomplishments became well-known in the region and even drew the attention of President Ulysses S. Grant, who hosted a special reception in their honor and shook hands with each of them.

Many of Connecticut's residents took a special interest in the Chinese Education Mission students, among them Samuel Clemens (Mark Twain) and the Reverend Joseph Hopkins Twichell. Speaking before an audience of Yale Law School students on a spring day in 1878, Twichell described a novel educational experiment that was going on in Connecticut's capital city, thirty miles to the north. "A visitor to the city of Hartford at the present time," he said, "will be likely to meet on the streets groups of Chinese boys in their native dress, though somewhat modified, and speaking their native tongue, yet seeming, withal, to be very much at home."[41]

Twichell became a friend to the Chinese youth and helped introduce them to American culture. He fed them venison and bear meat and took them on hikes (eight miles out, eight miles back) to an observation tower on a mountain near Hartford. The Clemens family likewise was active in befriending the boys, and held at least one neighborhood reception at their home.

The Chinese imperial government acknowledged the mission's success by appointing Chen Lan-pin and Yung Wing as China's first Minister and Associate Minister to the United States, respectively, and presented their credentials to President Rutherford B. Hayes. The two men effectively became China's first official diplomatic envoys to the United States.

The Chinese Education Commission erected its own building at 352 Collins Street in Hartford as a center for learning the Chinese classics. Hartford was chosen for the commission's headquarters for the city's central location and because Yung Wing had long been friends with educators at the Asylum for the Deaf and Dumb there, and he liked the area. There was another reason, however, according to Clara Day Capron, a child who lived in the neighborhood during the 1870s. Capron insisted, in an interview some eighty years later, that it was Yung's meeting and subsequent friendship with Twichell that finally convinced him to locate the headquarters there.[42] The decision made, Yung obtained authorization for the building's construction, at a cost of $75,000.

It was 1874, and the third detachment of thirty students was on its way (bringing the total number of students to ninety—the final thirty arrived two years later). The new facility (which was completed around 1877) would provide classrooms and boarding for seventy-five students, who would be required to gather at regular intervals to listen to the Chinese instructor read the Emperor's instructions. The curriculum of Chinese studies included the classics, poetry, calligraphy, and composition. For many of the newly Americanized Chinese students, however, the Chinese studies soon became a burden, and it was not uncommon to hear them referring to the Chinese Education

Commission building as "Hell House."⁴³ Unfortunately, their comments did not go unnoticed by the Chinese officials.

(As if Yung Wing did not have enough going on in Connecticut, in 1874, the Imperial Court asked Yung, who now had quasi-diplomatic status as a result of the mission exchange, to study the condition of Chinese workers in Peru. Yung invited Twichell and Yung's future brother-in-law, Dr. E. W. Kellogg, to accompany him. Outraged by the conditions he found, Yung Wing's subsequent report to the Chinese government helped put an end to Peru's infamous "coolie" trade.)

Meanwhile, back in Hartford, the students' distaste for Chinese studies was not the only indicator of mounting problems with the mission, in the view of the imperial government: The boys were adopting Western manners. Many had stopped wearing the traditional long Chinese gown, preferring Western suits, instead. Some had cut their hair, some even had converted to Christianity. A few were dating American girls and dancing at parties. Chen Lan-Pin, in particular, was shocked at the degree of Americanization among the students. He came to believe that the educational process was resulting in the alienation of young people from Chinese tradition and might create a cadre of "westernized" scholars who could pose a threat to the Confucian elite.

In 1880, a new commissioner, Wu Jiashin, arrived from China and was shocked at how Americanized the boys had become.⁴⁴ Huang Zunxian, documenting the unfolding events, wrote:

> Then came the new Principal, Mr. Wu,
> and his associate, who were fond of showing bureaucratic powers
> And who said, "These are runaway horses;
> We have to bridle them before we can ride them."
> ... The Principal rose up in a rage
> And panted like an air pump made of bamboo.⁴⁵

One of the mission students, Yung Kwai, later wrote: "When [Wu Jiashin] was appointed head commissioner, the fate of the mission was sealed.... What else could have been expected from a man brought up to shut his eyes to everything not Chinese for fear that his relish for the dry husks of Confucian classics might be spoiled thereby?"⁴⁶

> "A bird born in captivity can not indeed appreciate the sweet odor of the woods; but let it once have free space to exercise its wings, off it flies to where natural instinct leads."
> — Chinese Education Mission student Yung Kwai

At the same time, intolerance for minorities and immigrants was growing in the United States. After the period of acceptance that had followed the Civil War, the political climate was changing. Among other things, a sudden drop in the labor market following the completion of the transcontinental railroad led some Americans to revolt against Chinese laborers in Western states. Many Chinese workers were now without jobs and allegedly competing with American workers who also needed work. The earlier disposition for tolerance toward foreigners was deteriorating markedly. Anti-Chinese agitation in California in particular festered from a combination of economic difficulties and racism.

Coinciding with Wu's arrival, anti-Chinese sentiment in California was pressuring Congress to consider severely limiting immigration and perhaps denying the Chinese citizenship. In 1879, President Rutherford B. Hayes vetoed the first attempt to pass such legislation. "I knew the President would veto that infamous Chinese bill," Samuel Clemens later wrote. Twichell took the grievance to the American Missionary Association: "It is one of the most humiliating confessions that can be made, to say that these people cannot be granted room on our soil, with liberty and justice under our laws, with safety to ourselves."[47]

Distaste for the project was growing on both sides, and Yung Wing's plan was finally abandoned in 1881. Simply put, American intolerance and Confucian conservatism won out over education and progress:

> They [the students] had been allowed to enjoy more privileges than was good for them ... they imitated American students in athletics ... they played more than they studied ... they formed themselves into secret societies, both religious and political ... the sooner this educational enterprise was broken up and all the students recalled, the better it would be for China.[48]

Many eminent Americans, including former President U.S. Grant, President Porter of Yale University, and Samuel Clemens, petitioned the Chinese government not to withdraw the students. In a subsequent letter to China's Ministry of Foreign Affairs, Yale president Noah Porter and others expressed their "exceeding regret" that the young men had in effect been withdrawn, and that the Education Commission had been dissolved:

> The studies of which they [the students] have been deprived by their removal, would have been the bright flower and the ripened fruit of the

roots and stems which have been slowly reared under patient watering and tillage. We have given to them the same knowledge and culture that we give to our own children and citizens . . . we would respectfully urge that the reason for this decision should be reconsidered.[49]

At the time of the recall, only two students had graduated from the Sheffield Scientific School of Yale University. Twenty others were studying at Yale, four at Columbia, seven at the Massachusetts Institute of Technology, and five at the Rensselaer Polytechnic Institute; the rest were still in preparatory high schools.

Some of the students who were well-established in colleges were allowed to stay on, at first. But increasing pressure transferred by correspondence between China and Connecticut finally severed all the commission's ties between the United States and China. By August of 1881 almost all the students were ordered home, and the Chinese Education Mission came to an abrupt end, a mere nine years after the first students arrived. Letters between Chinese officials concerning the end of the mission (written in 1880 and 1881) are included in Ssu-yu Teng and John King Field's *China's Response to the West, A Document Survey 1839–1926*.[50] They clearly express the concern felt by the Imperial officials for China and the boys' welfare.

Political relations between China and America had deteriorated generally to such an extent that some of the students who returned became virtual prisoners in Shanghai. One of those students, Yan Phou Lee, managed to escape in 1883 and return to America. Re-enrolling at Yale as a sophomore in the class of 1887, he became a distinguished student who received prizes in English composition and declamation, and was elected to Phi Beta Kappa.

Although the Chinese Education Mission was short-lived, this initial group of students provided China with its first Western-trained technicians and engineers. Several of them went on to play important roles in modernizing China and in promoting education exchanges with the West.[51] One of the first boys sent by the commission, Tong Shao-yi, later became prime minister of the Chinese Republic. Another of the mission's students, Chan Tien-Yu, also made important contributions to his homeland. Chan was just eleven when he came to Connecticut with the first installment of students in 1872. He graduated from the Sheffield Scientific School of Yale University in 1881 and returned to China, where he dedicated the next thirty-two years of his life to the design, planning, and construction of China's railroads. One of them, the famous Peking-Kalgan Railway, was the first to be built solely by the talent of Chinese engineers and without any foreign assistance. Today, the name of Chan Tien-Yu is synonymous with the spirit of self-sufficiency on both sides of the Taiwan Straits, symbolizing the Chinese people's determination to modernize China.[52]

After his mission ended, Yung Wing spent time in the United States and China, advancing business and political ventures, until the Hundred Days' Reform of 1898. Because he publicly supported the progressive movement, he felt compelled to flee to Hong Kong when the Empress Dowager suppressed it. He was in for a surprise when he arrived at the U.S. Consulate and discovered that he was no longer considered a United States citizen. The Chinese Exclusion Act, which had become law in 1882, denied U.S. citizenship to people of Chinese birth—in spite of the fact that he had been granted citizenship in 1852. Yung returned to America nevertheless, where he remained for the rest of his life. He died in Hartford in 1912.

Before his death, Yung Wing made a final contribution to education exchange with America by donating his prized collection of Chinese books to Yale. The acquisition formed the nucleus of Yale University's Sterling East Asian book compilation, and initiated one of the finest collections of its kind in the West. Yung Wing's portrait hangs in Yale University Center, alongside those of other famous graduates.

Yung Wing's diary, published in 1912, contains many personal and touching accounts of his life and thoughts:

> All through my college course, especially in the closing year, the lamentable condition of China was before my mind constantly and weighed on my spirits. In my despondency, I often wished I had never been educated, as education had unmistakably enlarged my mental and moral horizon, and revealed to me responsibilities which the sealed eye of ignorance can never see, and sufferings and wrongs of humanity to which an uncultivated and callous nature can never be made sensitive. The more one knows the more he suffers and is consequently less happy. The less one knows, the less he suffers, and hence is more happy. But this is a low view of life, a cowardly feeling and unworthy of a being bearing the impress of divinity. I had started out to get an education. By dint of hard work and self-denial, I had finally secured the coveted prize and although it might not be so complete and symmetrical a thing as could be desired, yet I had come right up to the conventional standard and idea of a liberal education. I could, therefore, call myself an educated man and, as such, it behooved me to ask, "What am I going to do with my education?" Before the close of my last year in college I had already sketched out what I should do. I was determined that the

rising generation of China should enjoy the same educational advantages that I had enjoyed; that through western education China might be regenerated, become enlightened and powerful. To accomplish that object became the guiding star of my ambition.[53]

The First Enrollments from Japan

While the Chinese Education Mission was flourishing, Japan also began sending students to American colleges and universities.[54] A key figure in the initiation of the earliest Japanese enrollments was Julius H. Seelye, president of Amherst College (1876–1890), an ordained minister in the Dutch Reform Church and a member of the American Board of Foreign Missions. In 1871, Seelye invited members of Japan's diplomatic Iwakura Mission, who were seeking a system of education to support the building of a modern nation, to visit Amherst. The visit was reciprocated when Seelye went to Japan in 1872, where he dined with Emperor Meiji. Seelye boasted in his diary that he had also preached the first public sermon in Tokyo since the government had lifted its ban on Christianity. The visits served to establish a long and friendly association between Amherst and Japan, resulting in some of the first Japanese enrollments in American colleges.

One of those initial enrollees at Amherst was Niijima Jo, who would later come to be known as Joseph Hardy Neesima. Born of samurai parents, he was only ten years old when Commodore Perry's arrival altered Japan's isolationist policies, opening doors to Western thinking and religion. When Neesima was twenty-one, he became a Christian after reading an abridgement of the Bible, and made the decision to study theology in the United States. Driven by his convictions and at the risk of life and limb, he left Japan without governmental permission and ultimately reached the United States by working his passage on an American schooner, aptly named the *Wild Rover*. Upon arrival in Boston, the owner of the ship, Alpheus Hardy, agreed to sponsor the young student and enrolled him in Phillips Academy at Andover in 1865. Neesima went on to graduate from Amherst College in 1870 and then from Andover Theological Seminary in 1874. He was the first Japanese to obtain an academic degree in the United States.

Through a fortunate opportunity to act as an interpreter for a Japanese embassy in 1871, Neesima received a formal pardon from Japan for having left the country without permission. In addition to the pardon, he was granted the privilege of teaching Christianity upon his repatriation to his homeland. Returning to Japan in 1874, he was convinced of the necessity of establishing an institution of higher learning in his own country based on Western ideals and Christian moral teachings.[55] So, in 1875 he founded Doshisha Eigakko (Doshisha College) in the heart of Kyoto, right across from the imperial palace. It later became Doshisha University.

Throughout his career he encouraged the education of Japanese students in America, and Doshisha University remained active in student exchanges with the United States. By 1904, partly through the efforts of Doshisha, students from Japan accounted for 105 of the 2,673 foreign students studying in the United States, according to the United States Bureau of Education.

Other adventurous young Japanese followed Neesima to Amherst. Kanda Naibu (Class of 1879), aided by the Japanese government, which was now increasingly enamored of Western modernity, arrived in Amherst in 1871 at the age of fourteen. Seelye arranged for him to stay with an Amherst family and attend the local high school until he was ready for Amherst College, where he later enrolled. Like Niijima, Kanda devoted his life to Christianity and new education in Japan. After returning to his country he became Japan's preeminent authority on English and Latin, served as president of Japan's English Speaking Society, and worked to make the study of English an integral part of Japan's national education.

Another Seelye disciple, Uchimura Kanzo (Amherst Class of 1887), came to be known as the founder of "Japanese Christianity." Uchimura had been influenced by William S. Clark (president of Massachusetts University of Agriculture) in Hokkaido in the 1870s and had been recommended to Seelye by Niijima. While Uchimura was a student in America he was frequently questioned about his conversion to Christianity, prompting him to publish a book, *How I Became a Christian*. As time passed, however, he reportedly came to mistrust the influence of foreign missionaries, and began to claim in his writings that Christianity in fact had Asian origins.[56] Based on his assertions he founded what he called the "No Church" (mukyokai) movement, the first Japanese Christian sect. Uchimura advocated scholarly study of the Bible rather than organized church services, and he became a popular lecturer. In the 1920s, he was a sharp critic of the American government's exclusion policies that barred Japanese immigration. His followers included leading Christian intellectuals such as Tanaka Kotaro, a postwar minister of education and chief justice of the Japan's

Supreme Court, and Nambara Shigeru, the president of Tokyo University.

Another Amherst graduate who made a lasting impact in Japan was the son of Admiral and Count Kabayama, Sukenori Kabayama Aisuke. After receiving his degree in 1889, he served in government and in business before becoming president of the Japan-U.S. Friendship Society in Tokyo, where he befriended U.S. Ambassador Joseph Grew. That association, assisted by funding from John D. Rockefeller, led to the establishment of the International House of Japan many years later (in 1956), which encouraged intellectual exchange between nations. Their efforts also initiated the Grew Foundation, which provided fellowships to Japanese students for study in American colleges and universities.[57]

The Beginnings of Immigration Policy

America's immigration policies have affected foreign student enrollments from the beginning. Historically, regulations and rules employed to enforce or restrict immigration have reflected the political and social climate of the nation—as the country's mood shifted, so did its policies. In most cases, although foreign students were exempted from many of these restrictions, they still had to deal with many of the same logistical issues (and biases) as any other immigrant.

Before 1882 almost anyone could come to America. They simply disembarked from their ship or crossed the border and went about their business as they pleased, with little notice taken by the U.S. government.

That began to change during the 1830s, which saw growing concern over the increasing volume of immigration both in absolute terms and in terms relative to the U.S. population.[58] Growing numbers of the new arrivals at that time were Catholic and unskilled, partly as a result of the potato famine that had devastated Ireland and much of northwest Europe.

Development of the steamship also contributed to immigration growth. Before the 1840s, immigrants had most often arrived on sailing ships, and an average voyage from the British Isles to the United States lasted five to six weeks; a trip from the continent took a couple of weeks longer. During the 1850s, however, steamships became more common, and not long after the end of the Civil War, they became the preferred mode of travel. Just as importantly, steamships provided a much greater degree of comfort and safety than did sailing ships, making transatlantic or transpacific voyages far more palatable than they had been previously.

Prior to 1855, most immigrants, including foreign students, had arrived in America at any one of many points of entry. During the colonial and early national periods, immigrants typically arrived in New York City, Philadelphia, Boston, New Orleans, or Baltimore. After steamships took over the seagoing industry, however, New York City became the main arrival port, and it was here that America's first immigration facility, Castle Garden, was established. This complex, located at the tip of Manhattan, was replaced a few decades later, in 1892, by a much larger facility at Ellis Island.

Around 1870, the growing concern about the numbers of immigrants entering the country resulted in a handful of individual states passing restrictive immigration laws. To circumvent the potential confusion that state-by-state legislation would create, the United States Supreme Court declared in 1875 that immigration policy should be consolidated under federal control.[59] As the number of immigrants rose during the 1880s, Congress began to formulate immigration legislation, most of which directly targeted transplants from China. Citizens around the country were pressuring the government to restrict what was widely considered to be excessive Chinese immigration to the United States.

The discovery of gold in California in 1847 and the resulting "gold rush" had attracted fortune hunters from the world over, particularly from China. Most of the people who made the journey were poor, were uneducated, and had few skills. Not surprisingly, the overwhelming majority failed to strike it rich in California and, adding to their misfortune, most were left with no means of paying for passage back to China. Thus, thousands of Chinese were forced to remain in America and find work, often as underpaid laborers, in order to survive. The Chinese are credited for having provided much of the physical labor required in building America's first transcontinental railroad, which became the lifeblood of the nation's economy.

As was noted previously, when the railroads were completed and the legions of Chinese laborers were again forced to seek employment, American resentment of these workers grew. The "yellow peril" was blamed, among other things, for taking needed jobs from U.S. citizens. Partly as a means of self-protection against outraged Americans, many Chinese created small, close-knit communities in American cities; these areas became known as Chinatowns.[60] Even as the Chinese pulled together for support, violent outbreaks grew more frequent, often occurring within the Chinatowns themselves. American sentiment, at least among the more vocal citizens and lobbyists, had become decidedly anti-Chinese.

The government's response to the unrest was the passage of the Chinese Exclusion Act of 1882 that prohibited certain laborers from entering the United States. The legislation prevented any Chinese without family already in the U.S. from entering the country, thus effectively halting almost all new immigration from China. The law was the first in U.S. history to ban a specific ethnic group from entering America, and under its provisions only diplomats, merchants, and students were allowed access to the United States. Also in 1882, Congress imposed the Immigration Act, which was more far-reaching. It bolstered the Exclusion Act and levied a head tax of fifty cents on each immigrant likely to become a public charge.[61]

The enactment of the 1882 legislation coincided with the closing of Yung Wing's education mission. Americans' increasing intolerance toward immigrants, especially Chinese immigrants, during the 1870s fueled the growing distrust between China and the United States during those years. This undoubtedly contributed to the Imperial Court's decision to recall the mission's students in 1881.

With the division between the countries continuing to widen, the Scott Act of 1888 prohibited immigration of virtually all Chinese, including those who had gone back to China to visit and had planned to return. In 1889, the Supreme Court ruled in a Chinese exclusion case (*Chae Chan Ping v. United States*) that an entire ethnic group could be denied entry to the country if the government deemed it difficult to assimilate, regardless of prior treaty.[62]

The loss of foreign students was one of many repercussions of the new restrictions. Another was the immediate need to establish an enforcement agency. To this point, state boards or commissions had enforced immigration law with direction from United States Treasury Department officials. At the federal level, U.S. Customs Collectors at each port of entry collected the head tax from immigrants, while Chinese Inspectors enforced the Chinese Exclusion Act.

Congress soon expanded the list of excludable immigrants and consequently made enforcement even more complex. In 1891, the Immigration Act was updated to bar "polygamists, persons convicted of crimes of moral turpitude, and those suffering loathsome or contagious diseases" from entering the United States. At roughly the same time, Congress created the Office of the Superintendent of

Immigration. Housed at the Treasury Department, the Superintendent oversaw the first of the soon-to-be legions of U.S. immigration inspectors stationed at the nation's points of entry.[63]

The decade of the 1890s marked America's initial attempts to implement national immigration policy in a comprehensive way. In compliance with the new laws, the federal government took on the tasks of inspecting, admitting, rejecting, and processing all immigrants seeking admission to the United States. The new immigration service's first duty was to collect arrival manifests from incoming ships, previously a responsibility of the Customs Office.

Operations officially began at a new federal immigration station on Ellis Island in New York on January 2, 1892.[64] The new complex housed inspection facilities, hearing and detention rooms, hospitals, cafeterias, administrative offices, railroad ticket offices, and representatives of many immigrant aid societies. At the time no one could have predicted the unrelenting tide of immigrants to come. Between 1892 and 1953, more than twelve million immigrants were processed at this one facility.

At the time of its inception, Ellis Island employed 119 of the Office of Immigration's total staff of 180. That number would rapidly expand as additional immigrant stations soon appeared at other U.S. points of entry, such as Boston and Philadelphia. To finance the effort, an "immigrant fund" was created from the collection of the "head tax" which remained in effect until 1909, when Congress replaced the fund with an annual appropriation. Congress exerted further federal control over immigration with the Act of March 2, 1895, which upgraded the Office of Immigration to the Bureau of Immigration and changed the administrator's title from Superintendent to Commissioner-General of Immigration. The Act of June 6, 1890, consolidated immigration enforcement by assigning both Alien Contract Labor Law and Chinese exclusion responsibilities to the Commissioner-General.

Thus, in the span of not much more than a decade, the United States had moved from an open-door policy to one of selected exclusion. Those not excluded were required to provide more documentation and in some cases to endure complicated processing, all of which demanded a sizable staff and an impressive facility.

At the turn of the century, few could have accurately predicted the tide of immigrants who would be drawn to America. As the historian Frederick Rudolph later recorded, a progressive era was affecting the United States in the wake of urbanism and industrialization. America was becoming increasingly irresistible to hopeful immigrants, students, and visitors from every region of the world.

3

The Early 1900s: Foreign Student Enrollments and Emerging Support

The progressive movement of the late nineteenth and early twentieth centuries was one of several ideological forces that contributed to the rising status of American institutions of higher learning in the early 1900s. Progressivism was partly an expression of social conscience, generated to a large degree by the growing middle class, in the presence of conditions born of the urbanization and industrialization of a formerly agrarian society. Following decades of what historian Frederick Rudolph describes as "free-wheeling, atomistic individualism," a movement that elevated service as a motivation for learning began to gather strength:

> Progressivism was Theodore Roosevelt as a police commissioner of New York, setting forth in a black cloak at midnight, in search of crime and delinquent police officers ... it was Robert La Follette fighting the lumber interests of Wisconsin, as elsewhere good Progressives fought other interests of privilege: the railroads, the utility gang, the sugar trust, the farm-machinery trust, even the bicycle trust. Progressivism was a gigantic effort to deal with the discovery that the United States was a land of small farms and country stores no longer....[1]

The new social emphasis was on service, technology, and practical learning. As this perspective moved to the forefront of American awareness, it encouraged the development of a vast array of technical advances in business and industry. To the extent that higher education was enlisted in the work of social amelioration demanded by progressives, academe could not help but be profoundly affected. State universities became major teaching-training agencies bent on raising achievement standards for the nation's secondary schools. Engineering departments in the larger research-oriented institutions, for their part, took up the challenges of creating new machines of production, inventing innovative products, and devising solutions for complex industrial problems. Public land-grant schools likewise grew more active than ever before in developing and delivering agricultural services, especially those directly beneficial to the rural sector of the American economy.

Nonetheless, in the years leading up the new century, foreign student populations in U.S. colleges and universities remained modest. There is little doubt that American isolationism served as a major deterrent to the growth of the numbers of foreign students. Another obstacle was the reputation for academic excellence leading European universities continued to enjoy. Moreover, differences between American curricula and the academic requirements of other countries made an accurate evaluation of foreign students' credentials a logistical chore few admissions personnel knew how to handle. Social adjustment issues and barriers of language further complicated the difficulties involved. All told, at the turn of the century, support systems to meet the needs of nonnationals were virtually nonexistent.

In addition to these logistical, linguistic, and social issues, most American colleges in the opening year of the twentieth century were not especially receptive to the idea of educating foreign students. In one typical survey conducted in 1910, a large percentage of the schools queried expressed no more than mild interest in accepting students from abroad. For many educators, unattuned as they were to the idea of global education exchange, encouraging the enrollment of foreign students was an unfamiliar, even uncomfortable prospect.[2]

The Foreign Student Census 1900–1930

Prior to mid-century, when the Institute of International Education initiated a comprehensive annual census that defined specifically who should be counted as "foreign students" in United States higher education, enrollment figures were imprecise. Counts compiled by the Committee on Friendly Relations Among Foreign Students, and estimates provided by various annual editions of the World Student Christian Federation's publication, *Reports of Student Movements*, help to document the totals as best they could. The Bureau of Education also had been concerned with compiling an accurate census of foreign students since 1904. Discrepancies existed, however, among the agencies in determining exactly who was included in their respective counts. On occasion, high school students were included in the same totals as graduate-level students, and in some cases there was debate as to which institutions should be considered part of higher education.[3] Hence, foreign student statistics prior to 1949 are best considered approximate at best.

In its statistics record for 1904, the United States Bureau of Education reported a total of only 2,673 students from 74 countries in American institutions of higher learning (these figures excluded colleges for women only). Compare this to the total number of foreign students in Germany at the time, which was almost 9,000, while approximately 2,000 foreign students were attending colleges and universities in France.[4]

Of the total population of foreign students enrolled in U.S. collegiate programs in 1904, British North Americans (from Canada, Newfoundland, and so on) were the leading group, numbering 614. Mexicans were in second place, with 308; Cubans were third, with 236; the Japanese contributed 105; and the Chinese came in fifth, with a total of 93. The Philippines were also represented, with a total of 46 students on the 1904 roster.[5]

Indicative of no more than slow, incremental growth, in 1911–1912 the foreign student population was reported to have reached a total of 4,856, according to an August 28, 1912, United States Commerce Report. The leading country of origin was Canada, with 898 students, followed by the West Indies, with 698. China sent 549 students; Japan, 415; Mexico, 298; the United Kingdom, 251; and Germany, 143. India and Ceylon combined sent 148 students; Russia and Finland together sent 120. Brazil sent 76; Argentina, 51; Peru, 28; Colombia, 28; Chile, 19; and the other South American countries, 72.

In 1920–1921, the Department of the Interior, which oversaw the United States Bureau of Education, fixed the number of foreign students in American colleges and universities at 8,357. The highest representation was from China, with 1,443. Canada came in second, with 1,294. Other significant representations were from the Philippines, 857; Puerto Rico, 302; Japan, 525; Russia, 291; Mexico, 282; India, 235; Africa, 223; France, 160; and Great Britain, 149.[6]

Between 1900 and 1930, foreign student enrollments in the United States rose steadily. The numbers during this time were at last sufficient for institutions to take notice. Foreign student migrations were proceeding from new directions, bringing more enrollments from Latin

American countries and Europe, and a return of Chinese students after a long absence. In response to the increasing evidence of a foreign-student presence on America's campuses, universities created and developed an array of organizations, services, and policies to support and direct the foreign student experience in American higher education.

In the Spirit of de Miranda: Growing Latin American Enrollments

Francisco de Miranda's admission to Yale in 1749 marked a long and enduring relationship between Latin American students and American higher education. Enrollments were few throughout the eighteenth and nineteenth centuries, however, and remained so until around 1903.

Nevertheless, after a long period of internal conflict and revolution in South and Central America near the end of the nineteenth century, a new era of peace in the early twentieth century prompted a fresh emphasis on both education and travel. Peace in Latin America generated new prospects for economic growth and attracted foreign capital. Railways were built, and transportation was markedly improved between cities and into previously isolated areas. With greater access and exchange came increased commerce. An expanded and diversified economy demanded more skilled workers, such as people trained in running businesses, as well as engineers, educators, politicians, and specialists and professionals of all descriptions.

In many cases, Latin American students chose to acquire the skills needed to fill those positions in the United States, which seemed, in the eyes of much of the world at the time, a land of fresh possibilities. And, new transportation systems made it increasingly easy and relatively inexpensive to get there. Another factor influencing the increase of Latin American students in U.S. institutions was the fact that English had begun to displace the traditional French in the Latin American public school curriculum. Given their increasing ability to communicate in English, many more Latin American graduates, particularly Colombians, were drawn to U.S. institutions.[7]

Many of the institutions that initially attracted Latin American students, particularly those from Mexico and Central America, were sectarian: the Catholic schools in California, the College of the Christian Brothers in St. Louis, and the Convents of the Sacred Heart in St. Charles, Missouri, to name a few. Connected by faith and focus, many Latin American students chose to study in Christian institutions, where they could be educated by knowledgeable instructors, mentors who shared and understood their religion and its application to their education goals.

Governmental support from a number of Latin American countries also contributed to the growing migration of students to the United States. Venezuela, for example, was one of the first Latin American countries to appropriate scholarship funds for research in foreign lands (1909), and most of the scholarship's recipients, headed straight for the United States, partly because of its proximity.[8] Other South American countries took similar steps, and combined with the region's expanding commerce and transportation, by 1921 the number of Latin American students attending American institutions of higher education had reached nearly 5,000. The majority was enrolled in colleges, universities and technological schools, and the remainder in preparatory schools. At the time, the top field-of-choice for Latin American students was engineering. Many students also were enrolled in liberal arts programs, followed by medicine, dentistry, and commerce.

The Boxer Rebellion and the Return of the Chinese

The early 1900s saw a sudden resurgence of Chinese students in the United States—the indirect result of a bloody conflict inspired by the Imperial Court that came to be known as the Boxer Rebellion. When China was defeated by Japan in 1895, European powers responded with a scramble for so-called "spheres of interest," which primarily involved holding leases for railway and other commercial privileges in various regions of the country. The Russians took control of Port Arthur; the British acquired the New Territories around Hong Kong; and the Germans were granted a leasehold in Shantung. The American government supported an "open door" policy in China through which many commercial and trade opportunities became available.

An already enfeebled Imperial Court viewed all such incursions as a form of occupation and responded by quietly funneling aid to various secret and "underground" societies in an effort to covertly undermine the foreign intruders. Traditionally, and ironically, these societies had been formed to oppose the Imperial government. Antiforeign sentiment became so pronounced in China, however, that at the century's end the reigning Empress Dowager hoped that the secret societies would work with the government toward a common goal: removing the outsiders. The Empress began covert negotiations with a secret society called the Boxers.

The Boxers, or "Righteous and Harmonious Fists," were a religious group that had originally rebelled against the Imperial government in Shantung in 1898. They practiced animistic rituals and cast spells that they believed made them impervious to bullets and pain. The Boxers also believed that the expulsion of "foreign devils" would magically renew Chinese society and begin a new golden age. Much of their discontent, however, was focused on the hardships of the 1890s, which they blamed almost entirely on the government. Now, however, the Boxers and the government had a common enemy. With a covert nod from the Empress, the Boxers turned their aggression toward the unwelcome Westerners.[9]

The actual rebellion was limited to a few places, concentrating itself in Peking (Beijing). The Boxers attacked Western missionaries and merchants as well as compounds where foreigners lived, beginning a siege that lasted eight weeks. Horrified victims and witnesses to the assault scrambled to send urgent messages to the West for help and rescue.

In response to the outcry, 19,000 troops of the allied armies of the West were sent to end the siege. On August 14, the allies captured Peking. The Europeans were rescued and the Imperial government was forced to submit to the draconian terms of the Boxer Protocol of 1901. Under the protocol's provisions, Western powers now had the right to maintain military forces in the Chinese capital, thereby placing the imperial government under what amounted to a form of continuous arrest. The protocol suspended the civil service examination, among other things, and several high-ranking Chinese officials were prosecuted for their roles in the rebellion. To make matters worse, China also was ordered to pay a huge indemnity for the human and monetary losses, which included almost $25 million to the United States.[10]

In addition, the conditions of the Boxer Protocol served in part to set China on a new course of political and educational reforms. In 1901, the education system was restructured to allow the admission of girls, and the curriculum was changed from the study of the classics and Confucian theory to the study of Western mathematics, science, engineering, and geography.

United States President Theodore Roosevelt was eager to encourage China's new direction. In 1908 he put forth a plan to forgive a significant portion of the Boxer Indemnity owed to the United States by China. The revocation of the debt was presented to the imperial government with the suggestion that a good portion of the returned money be used for continued education reform—including exchange with the United States. Roosevelt defended the decision to his American constituency:

> This nation should help in a very practicable way in the education of the Chinese people so that the vast and populous Empire of China may gradually adapt itself to modern conditions. One way of doing this is by promoting the coming of Chinese students to this country and making it attractive to them to take courses at our universities and higher education institutions. Our educators should, as far as possible, take concerted action toward this end.[11]

It had been forty years since the hopes of Yung Wing and the Chinese Education Mission had been thwarted by Confucian reactionaries. But now, partly as a result of the remission of the Boxer Indemnity funds, China was motivated once more to send students to America. In 1899, only approximately eighty Chinese were studying in American colleges, but by 1912, with the help of governmental support and indemnity funds, the count jumped to nearly 800. In 1914, China also began to include women among the students sent to the United States.

One of the groups charged with the distribution of the proceeds of the Boxer Indemnity was the Board of Trustees of the China Foundation for the Promotion of Education and Culture, which consisted of ten Chinese and five Americans.[12] It was through the board's effort that Chinese student recipients moved successfully through the process of getting safely to America, enrolling, and graduating. The board was comprised of a selection of well-known educators and dignitaries. One of their meetings was attended by W. T. Tsur, former president of Tsing Hua University; V. K. Wellington Koo, Minister of Foreign Affairs; Chang Poling, President of Nan Kai University, Tientsin; W. W. Yen, Premier of China and Chairman of the Board; and Fan Yuan Lian, President of the National Normal University, Peking. Nearly every Chinese member of the Board of Trustees had at one time studied in the United States.

A 1909 Foreign Student Services Survey

Chinese student enrollments rapidly increased throughout the United States during the early twentieth century, a fact that was not lost on educators who had inherited new problems and issues surrounding foreign student admission, language, learning, and social acclimation. While institutions such as Yale had a long history of working with foreign enrollees, many colleges and universities in the United States were poorly equipped to serve the needs of nonnationals. A general consensus of opinion held that attention should be paid to providing help and service to this growing population. In 1909, to ascertain how United States higher education was providing for Chinese students, the Bureau of Education circulated a letter to about 100 institutions, asking the following questions:

1. Have students from China at any time attended the institution under your direction?
2. Would proof of equivalent attainments, including a preliminary knowledge of English, be accepted in the case of each student, in lieu of the usual admissions requirements or examinations?
3. What special inducements does your institution offer in the way of scholarships available for Chinese students, or fraternities or societies open to them?
4. What facilities in the way of summer sessions or special courses of study for such students are possible under your plan of organization?
5. What would be approximately the annual expense to each student—for tuition; for living expenses?
6. What residence accommodations could be secured by Chinese students during the terms and during vacation?
7. What special care would be exercised over the welfare of such students?
8. Other information that may occur to you as bearing upon the subject, including the title of any handbook issued by your institution giving fuller details of interest to Chinese students.[13]

The questionnaire was circulated among a broad cross section of institutions of varying types and sizes. The general state of Chinese student enrollments, costs, and support in the United States just after the turn of the century can be guessed at by a brief summary review of the responses from several of the institutions that participated in the survey.

One of the respondents, for example, was the University of California at Berkeley, which reported a total of nine Chinese students enrolled for the year 1907–1908, and 19 registered for the 1908 summer session. Tuition cost $20 annually, and living expenses were estimated to be about $400 per year. Most of the Chinese students at Berkeley resided with private families rather than on campus, although a small clubhouse reserved for Chinese students was home for a few. The institution's department of oriental languages was charged with their special care and guidance.

Yale University returned the following narrative response to the questionnaire:

> We have had a large number of students from China at Yale. During this past year (1907–1908) there has been an enrollment of about 25. Among the men of distinction who have graduated in the past years have been Jeme Tein Yow, of Peking, and Yu Chuan Chang, who passed first in the great examinations in Peking two years ago. A list of Yale men in China was published some few years ago.... The only change that we make in entrance requirements in the case of Chinese students is that a knowledge of the Chinese language and literature is accepted in place of the Greek requirement of its alternatives in the academical department, and the substitution of Chinese for Latin in the scientific school. We make a special point of emphasizing the importance of a good knowledge of English before admission. Degrees from representative Chinese institutions, such as St. John's College, Shanghai, and Tientsin University, are accepted for admission to the graduate school, as would be the degrees of American institutions of rank.... We have at present two Williams scholarships at $100 a year. In addition, special scholarships have been awarded from time to time. For instance, a few years ago one graduate of the university stated that he would be responsible for meeting the tuition charges of ten Chinese students. There is a special organization of Chinese students. There is also a Cosmopolitan Club which has members from all the different nationalities represented at the university.... We have no special summer session that would be of use to Chinese students.
>
> The cost of tuition in the various departments is from $155 down. A large number of scholarships and fellowships are available to students of character and capacity. Good board can be secured from the university for about $4 a week. Rooms vary in price; I presume that most of the Chinese students pay about $2 per week for a room. Chinese students would be admitted to university dormitories on the same conditions as all other students. There is a distinct esprit de corps among the Chinese students at the university, and probably a rather unusual interest in China and the Chinese. This is partly due to the fact that Yale graduates have founded a collegiate school which they

hope to develop later into a university, at Changsha, the capital of the province of Hunan.[14]

Indiana University at Bloomington sent a much shorter response, as they had only one Chinese student enrolled, who lived in a private home off campus. The school's annual tuition was reported to be $40 and living expenses were estimated at $300 per year.

Many large institutions, such as the State University of Iowa at Iowa City, the University of Minnesota at Minneapolis, and Baylor University at Waco, Texas, reported having no Chinese enrollees. Most of these, and other, institutions did have some support in place, such as faculty advisors and scholarships, in preparation for anticipated foreign enrollments.

Wellesley College in Massachusetts reported only one Chinese enrollee, although three scholarships had been established by the college at the time of the visit from Chinese commissioners in 1906. Foreign students were housed off-campus in the town of Wellesley, and a faculty member acted as advisor. Tuition at Wellesley was $175 annually, expensive in comparison to most institutions, and living costs were estimated to be $275 per year.[15]

It is evident from the responses that there was little uniformity or definition relating to the education or support of Chinese students in colleges or universities in 1909. Admissions standards, English requirements, and living conditions varied dramatically from college to college. Anticipating the continuing rise in Chinese student enrollment, however, many institutions were beginning to put people and systems into place to help meet their needs.

Tsing Hua College

The success of their students in the United States was important to the Chinese, too. By 1917, the influence of Chinese American-educated citizens on Chinese society was becoming so important that Tsing Hua (Qing Hua; Qinghua) College in Peking published a *Who's Who of American Returned Students*. It contained extensive documentation of biographical records of Chinese students who had studied in America, and the importance of the positions they held after they returned to China.

Tsing Hua College, located in the northwestern suburbs of Peking (Beijing), had been established in 1911 on the site of the Qing Hua

Yuan—a former royal garden of the Qing Dynasty. Funded by the Boxer Indemnity, it had first been established as a preparatory school called Tsinghua Xuetang for those students who were sent by the Chinese government to study in the United States. In 1912 it became a college.[16]

Tang Guoan, a native of Guangzhou province, served as president for its first two years, from 1912 to1913. Tang was himself a product of education exchange. He had been sent to the United States in 1873—one of the first students chosen by Zen Guofan to go abroad under the auspices of the Chinese Education Mission—and had studied at Yale University. During his brief presidency, he selected fifty-nine Chinese high school students to further their education in America, thereby initiating a long tradition of education exchange between Tsing Hua College and United States higher education.[17]

By 1921, through the efforts of Tsing Hua College and other Chinese institutions and organizations, there were 1,218 Chinese students enrolled in American colleges and universities, of whom 65 were women. Most were enrolled in liberal arts programs; the second-highest number studied engineering. Other top choices of study for the Chinese were commerce, agriculture, education, medicine, and theology.[18]

Teachers College at Columbia University

Chinese students in American colleges and universities helped generate a new wave of influential scholars. The International Institute of Teachers College, Columbia University, was the first American institution to demonstrate on a large scale how an American college could adapt its program to fit the needs of foreign students. Under the directorship of Paul Monroe, orientation and guidance were provided for foreign students in terms of course selection and programs best suited their individual needs. Special classes in English were offered, and courses were altered in content and method to provide the best possible learning environment for foreign students. Supplemental educational experiences, such as field trips to American historical sites and other places of interest, were arranged for the foreign students. At one point during the International Institute's years of operation there were 3,852 foreign students, from eighty countries, enrolled in Teachers College.[19]

The institute had been made possible through a $1 million grant from the International Education Board, a group founded by John D. Rockefeller Jr. in 1923. The International Institute was designed:

1. to give special assistance and guidance to the increasing body of foreign students enrolled in Teachers College;
2. to conduct investigations into educational conditions, movements, and tendencies in foreign countries; and
3. to make the results of such investigations available to students of education in the United States and elsewhere in the hope that such pooling of information would help to advance the cause of education.[20]

Throughout the years of its operation (1923–1938), the institute produced an exceptional array of important Chinese graduates. One was Tao Xingzhi, who studied at Teachers College at Columbia during 1916–1917 under John Dewey, renowned philosopher and educational theorist. Tao returned home to become an influential proponent of modern education in China, and one of the most renowned educators in Chinese history. He developed an original synthesis of Deweyan and Chinese approaches to progressive education based on firsthand study and analysis of Chinese life and society. Tao Xingzhi was instrumental in putting together a series of educational surveys in Peking, Tianjin, and Shanghai in 1921 to determine the state of literacy. The study revealed an illiteracy rate of nearly 70 percent in some urban areas. Alarmed by the magnitude of the problem, Tao made the decision to devote most of his career helping to correct what he perceived to be one of China's biggest obstacles to progress.[21]

Hu Shih (Hu Shi) was another influential Chinese thinker who was American-educated. Winning a Boxer Indemnity scholarship in 1910, Hu enrolled at Cornell, and then later studied in the Department of Philosophy at Columbia University, also under John Dewey. Partly through Dewey's influence, Hu Shih was led to the conviction that China's modernization had to take place by changing ways of thinking and by writing Chinese, not in the classical style, but as it was spoken. After his return to China in 1917, Hu accepted an appointment at Beijing University and came to be regarded as one of the important intellectuals of his time. Advocating reforms on social, political, and intellectual issues, his was an influential voice in discussions about reevaluating Chinese culture. The Sino-Japanese War of July 1937 temporarily delayed his scholarly work, when he accepted a post as ambassador to the United States, where he served until 1942. He returned to China as chancellor of Beijing University in 1946. In 1949 the advance of Communism forced Hu Shih to flee to the United States, and he remained there until 1958, when he went to Taipei as president of the prestigious Academia Sinica research institute. Because of his contributions and legacy to Chinese culture, he is sometimes referred to as the "Father of the Chinese Literary Revolution."[22]

Chen Heqin (1892–1982) studied at Teachers College during 1917 and 1918. After graduation, Chen returned to China to become the first modern Chinese theoretician of early childhood education. His work included promoting opportunities for early childhood education, and he developed teacher-training programs that emphasized child psychology, family education, and education for handicapped children.

One of the first women to be granted an indemnity scholarship was Chen Hengzhe (Sophia Chen), who enrolled at Vassar College in 1914.[23] Chen helped to arouse interest in Chinese poetry through her articles in *Vassar Miscellany Monthly* and as a contributing editor to *Youth*, a magazine of contemporary poetry published by students at Cambridge.

"Slave Education"

The seemingly obvious benefit of Chinese students returning home with American degrees seemed incontestable to most people in the United States, yet that presumption was not universally embraced by the Chinese. In fact, the merits of an American education were vehemently rejected by some who felt that Western education simply could not effectively find practical use in China.

According to an article written by Chiang Yung-chen (then a professor at DePauw University in Indiana), disdain for foreign education spread across the ideological spectrum in China. From the 1920s on, earning Western degrees was a target of both public and academic criticism, both in China and the United States, and was but one element in an overall concern about China's modernization.

It was not just the Chinese who had their doubts about the value of Western degrees. American liberals such as John Dewey and Bertrand Russell, who had both taught in China, levied criticisms against modern Chinese education in general, and study abroad in particular.

Similar disdain was openly voiced by other notables such as Thomas Read of Beijing University; Robert McElroy, the first exchange professor at Tsing Hua University; and Selskar M. Gunn, vice-president of the Rockefeller Foundation.[24]

Criticism usually centered on two issues. The first was the problem of "blind copying," which Ching Yung-chen referred to as the "random grafting of Western twigs onto Chinese trees of which the Westerners had little knowledge."[25] In other words, excessive reliance on foreign theories and methods. What the returned students had learned in foreign countries was not always applicable to China.

Nathaniel Peffer, a journalist who reported from Shanghai and Beijing, was one of the harshest critics:

> From what I saw and heard in China during the five years I stayed there, I have to conclude that as a class the returned scholars were not only a sad disappointment as representatives of their nation but were also not successful despite having favorable opportunities. Now that I have returned to America and met with some Chinese students here, I find that their worst faults are their glib tongues, the mere lip service they give to patriotism, and the superficiality of their views.[26]

A compelling contribution to the criticism of Western education for Chinese students was *Jindai Zhongguo Liuxueshi* (*History of Studying Abroad in Modern China*), which was written by Shu Xincheng in the 1920s. Shu accused the returned students of becoming a sort of "special" class. He further criticized the students for taking on too many of the characteristics of foreigners and for forming factions promoting self-interest—the ultimate goal to seize power. These were the same accusations that had prompted the recall of the Chinese Education Mission students in the 1880s and, according to Shu, the criticisms were as pertinent as ever.

Shu was critical of the students, but his most disparaging accusations were saved for the Chinese government. Since the Qing dynasty, he claimed, when the first students were sent abroad, the Chinese government had no definitive purpose or policy regarding the practice. Further, few screening methods or uniform procedures were established. But the most serious governmental problems were the overly generous stipulations governing self-supported students and the exemption of these students from any screening examinations. This led to a waste of social resources, Shu claimed, and encouraged students to go abroad for "empty documents" rather than practical knowledge.

Shu held that the concentration of Chinese students in Japan and the United States created a special political and social status for the returned scholars that further contributed to the "Japanization" and "Americanization" of Chinese society. Moreover, the quotas of government-funded students sent abroad were unevenly distributed. Somehow, students from the provinces of Jiangsu, Zhejiang, and Guangdong always seemed to top the list. The social impact of that imbalance was a potential source of contention among the provinces and could be far-reaching.[27]

Wang Yiju's *Chinese Intellectuals and the West* extended Shu's argument, applying terms such as "slave education" and "worshipping and fawning on foreign powers"[28] Wang also traced the weaknesses of the returned Chinese scholars back to the groups first led by Yung Wing to America. He believed that Yung Wing had deliberately planned from the beginning to disregard the demands of the Qing court. Yung Wing was convinced, Wang believed, that China's future depended on these students abandoning their traditions, fully taking on the science and technology of Western culture, as soon as possible. That presumption had effectively betrayed the Chinese government's original purpose of assimilating and using Western technology while preserving traditional culture. The resulting premature recall of the students had, according to Wang, resulted in the mission being a poor investment for China.

The arguments on all sides were largely ideologically based. Confucianism and tradition stood in sharp contrast with American philosophies, and in the view of many observers, the training Chinese students had gained in America was not necessarily appropriate for China's future.

How American Friends Understand Us:

1. That Chinese are great eaters; their most favorable dish is the boiled rats.
2. That the Chinese eat opium . . . to the same extent . . . as Americans eat candy.
3. That the great industry of China is hand laundry.
4. That the Chinese never walk straight; they jump like the crows.
5. That most of the Chinese students in [America] are princes or sons of princes.
6. That the Chinese wear embroidered red gowns in their everyday life.

> 7. That men never make love to women; they are so modest!
> 8. That a man has at least 10 or 12 children.
> 9. That all Chinese officials are corrupt.
> 10. That a Chinese student who speaks English [attended a] missionary school.
> 11. That Confucius has a queue.
>
> **How We Understand American Friends:**
>
> 1. That Americans love dollars first and foremost.
> 2. That they are well-versed in the art of "bluffing."
> 3. That anyone can do anything in America if he can "pass the buck."
> 4. That they like to be flattered; they like to be called a sport.
> 5. That the gentle sex never takes life seriously; what they want is a good-time-right-now.
> 6. That they do not like to see a tragedy; the happy ending is absolutely necessary in any form of fiction.
> 7. That you can never tell when an American will laugh; he laughs at the most unexpected moment.
>
> Shen Hung, *Chinese Students Monthly*, January 1920, p. 59.

The Pensionados

From the late sixteenth century until the end of the nineteenth, the Philippine Islands were controlled by Spain. During this long occupation, most Filipino people had been converted to Catholicism—an important component of Spanish colonial rule. At the end of the nineteenth century, Spain ceded the Philippines to the United States after the Treaty of Paris ended the Spanish-American War, and in 1902, the United States passed the Philippine Organic Act, setting terms for civil government and endorsing Filipino self-government.

Also in 1902, American education was introduced to the Philippines and the English language was mandated for regular use in the school systems. To help integrate the new systems, the United States exported teachers—hundreds of them—to the Philippines.[29] In 1901, the "Thomasites," a group of 500 American educators, crossed the Pacific to the Philippines on the *USS Thomas* (thus the group's nickname) with the goal of educating Filipinos in the "American way" and establishing a new public school system for the island population.[30]

After being quarantined for two days after their arrival on August 21, 1901, the Thomasites were finally allowed to disembark from the ship. Initial assignments included dispatched the American teachers to the provinces of Albay, Catanduanes, Camarines Norte, Camarines Sur, and Sorsogon, among many others. The Thomasites were credited for preparing many of the *pensionado* students (i.e., scholarship recipients) for college in America.[31]

The *pensionado* program was adopted in 1903. Conceived at least in part by W. A. Sutherland, who had been sent to the Philippines the year before as an assistant to William Howard Taft, the program offered promising students scholarships to attend universities in the United States. The program provided government scholarships to students (*pensionados*) supposedly chosen by merit from each Philippine province, but in reality local prominence and connections played a major role in the selection process. In return for each year of education in the United States, *pensionados* were required to work for the government in the Philippines for the same length of time.[32]

Writing about the program at the time was L. T. Ruiz of Yale University, who had also served as secretary for Filipino students for the Committee on Friendly Relations. The Philippines had adopted a policy of sending some of their best-qualified students to the United States to study at government expense, according to Ruiz.[33] On October 13, 1903, the first group of 100 *pensionado* students sailed for the United States. The students spent a year (1903–1904) studying in high school in California, which allowed them to adjust to life in the United States and learn English. The *pensionados* then were dispersed to colleges and universities, especially on the East Coast and in the Midwest. These first Filipino students established an ongoing association between education and the state of Illinois, especially in the Chicago area.

Unlike the Filipinos who had been recruited for labor during the opening years of the century, the first Filipinos to arrive in Illinois came as college students.[34] A pattern developed naturally from the presence of so many Filipinos students in one location during the first decade of the century—42 of the 178 *pensionados* enrolled in U.S. institutions of higher education in January 1906 studied in Illinois. (Even after the abandonment of the *pensionado* program, young Filipinos associated Chicago with a community of fellow students. By contrast, they associated the West Coast, Alaska, and Hawaii with Filipino workers, who labored in fields and canneries or on plantations.) Others enrolled at the State Normal School in San Diego, the Berkeley, Cornell, Notre Dame, Purdue, Yale, Wisconsin, Northwestern, Nebraska, Colorado, Illinois, Indiana, George Washington, Iowa, Ohio State, Michigan State, and the Massachusetts Institute of Technology, among others. Some attended technical and vocational schools, while a few first enrolled in high school.

Later in 1903, three more students were selected to join the first allotment, bringing the total to 103. In 1904, another forty-three joined them, and the next year, thirty-five more were selected for the trip. The government program paid for the students' college tuition and fees, room and board, books, medical care, clothing, and laundry. The students also received an allowance of $70 per month ($80 if the student resided in New York City).

Of the more than 200 students sent by the program, only eight were women. The mark some of them made, however, was significant. One of those eight was Honoria Acosta-Sison, who would become the Philippines' first female physician. Rebelling against traditional Filipino culture, Honoria applied for a government scholarship as a *pensionado* in the first year of the program's operation and was accepted. She took preparatory courses first at Drexel Institute and at Brown Preparatory School, and then earned her medical degree from the Woman's Medical College of Pennsylvania. University of Pennsylvania professor Narciso Cordero, in his book *To While Away an Idle Hour*, writes about Dr. Acosta-Sison:

> She is the first Filipino woman physician, first Filipino woman graduate of an American Medical School, first Filipino woman obstetrician, and many other "firsts." To all these she reacts with her characteristic modesty, "Why all the fuss? Do you get excited simply because the fellow you have just met happens to be the first born in the family?"[35]

Olivia Salamanca, another of the eight female *pensionados* and the Philippines' second female doctor, followed in her predecessor's footsteps by graduating at the Woman's Medical College of Pennsylvania. There is little documentation of Dr. Olivia Salamanca's life, in part because she died very young of tuberculosis. The musician Juan Felipe (some suggest he was smitten with the young doctor) wrote a musical piece titled "Olivia Salamanca."[36] Today, at the intersection of Taft Avenue and T. M. Kalaw in Manila, one can find Olivia Salamanca Park and Dr. Olivia Salamanca Memorial Hospital located in Cavite City.

Fewer *pensionado* students were sent to America between 1905 and 1912, partly because the University of the Philippines was established in 1911. Also, the act that had provided for the maintenance of the students sent to the United States had been amended to apply only to those in graduate programs. As a result, the number of Filipino students sponsored by the government dwindled to only a few each year between 1912 and 1919.

That decline coincided with a time of increasing need for well-trained persons to command technical and scientific positions in the Philippines. An effort to fill the growing professional void prompted the Philippine legislature in 1918 to authorize additional funding to again send *pensionados* to the United States. By 1919, another 114 government-sponsored students had been sent to earn professional degrees. With rare exception, the later allotments of students were people who had been employed by the Filipino government for a number of years, and who had a special field of interest they were seeking to further develop. In 1922 the Bureau of Insular Affairs reported a total of 1,156 Filipino students in the United States. Categorized by field of study, 280 of those students were enrolled in engineering, 181 in commerce, 149 in medicine, 98 were studying law, 50 were enrolled in agriculture programs, 43 in education, and another 43 in the sciences.[37]

A number of important and historical figures in the Philippines were educated with the help of the *pensionado* program. Conrado F. Benitez was one example—sportsman, journalist, constitutionalist, lawyer, educator, civic leader, and economist.[38] He was born in Pagsanjan, Laguna, on November 26, 1889, the fourth son of Judge Higinio Benitez and Soledad Francia. As a boy, Conrado used to be called "Izutony Tubig" because he was constantly swimming in the nearby Pagsanjan River.

In 1911, Benitez was sent as a government *pensionado* to the University of Chicago, where he earned an MA and PhD. He was also the first Filipino to be elected captain of the university water polo team. He followed that achievement by winning three gold medals in the Far East Swimming Olympics held in Manila in 1913.

Once his swimming career was behind him, Benitez spent a number of years in higher education, serving in 1918 as head of the department of economics for the University of the Philippines. He later was appointed dean of the College of Liberal Arts, and was cofounder and dean of the University of the Philippines College of Business Administration.

Benitez also loved journalism. Following his resignation from the University of the Philippines, he accepted the editorship of the *Citizen*, the first Filipino weekly paper in English, then was later chosen as the first Filipino to serve as editor of the *Philippines Herald*. For these accomplishments, he is regarded by many as the father of Filipino journalism in English.

Benitez was one of the founders of the Philippine Columbian Association. As a delegate to the Constitutional Convention convened in 1935, Benitez was chosen as one of the "seven Wise Men" entrusted with the task of writing the final draft of the Constitution of the Philippines. He was responsible for introducing into the Constitution the idea of conserving the nation's natural resources and a provision for the general education of Filipino youth.

Another famous *pensionado* graduate was Camilo Osias, a Resident Commissioner from the Philippine Islands.[39] He attended the University of Chicago in 1906 and 1907, and graduated from Columbia University's Teachers College in 1910. He returned to the Philippines to teach, and became the first Filipino superintendent of schools in 1915 and 1916, and then served as an assistant director of education from 1917 to 1921. He was a member of the first Philippine mission to the United States in 1919 and 1920, and president of the National University between 1921 and 1936.

A well-known and respected citizen, Osias turned to politics and was elected a member of the Philippine Senate in 1925. He later served as Minister of Education of the Republic of the Philippines until 1945; as the Philippine representative to the Inter-parliamentary Union in Rome; and to the International Trade Conference in Genoa in 1948. He was an unsuccessful candidate for the Nationalist Party nomination for President of the Philippines in 1953.

Enrollments from Other World Regions:
France, Japan, and India

China, Latin America, and the Philippines sent the most students to the United States during the first part of the twentieth century, but other important exchange initiatives in that period drew students from additional countries as well. During World War I, for example, the Association of American Colleges developed the "program of international reciprocity." Under this initiative, the American Council of Education, the United States Office of Education, the French Ministere de l'instruction publique, and individual educational institutions in France worked together to send French students, particularly prospective teachers, to study in America. As a result of the combined effort, more than sixty-five colleges in twenty-three states offered scholarships to French women. The scholarships covered tuition and board, and in some instances also covered travel expenses and incidentals, for which the French government had appropriated 75,000 francs. By 1918, 113 French women seeking teaching degrees had accepted the scholarships and arrived in the United States. The next year they would be joined by eighty-seven more.[40] The program also offered scholarships to disabled French soldiers, and in the 1918–1919 academic year thirty-eight enrolled in American colleges. The Program of Reciprocity granted a total of 230 such scholarships, chiefly to women and military veterans. In 1920, the Association of American Colleges turned over the administration of the scholarship program to the American Council on Education.

Larger enrollments were coming to America from Japan, too. Katsuji Kato, secretary for Japanese students of the Committee on Friendly Relations, reported in the April 1917 issue of *The Student World* that since the first enrollment a half-century earlier, more than two thousand Japanese students had studied in the United States. A directory of Japanese students in North America in 1921–1922 showed an enrollment total of 865, 19 of whom were studying in Canada.[41]

Most Japanese students in the United States in the early twentieth century were pursuing degrees in theology; engineering and commerce were the second and third most popular fields, respectively. These students came chiefly from the Japanese middle class and had fathers who were merchants, officials, educators, or pastors. Among those Japanese students who enrolled in Pacific Coast institutions, however, larger proportions were from farming backgrounds, and many were of samurai lineage.

Although fewer in number than the Japanese, foreign student enrollments from India were beginning to increase in the early 1900s. During the early part of the century it was not common for Indian students to come to America. They preferred to go to England to study law or to sit for the Indian Civil Service Examination (ICS) for a lucrative job under the British Raj. According to some sources there were approximately 100 Indian students in the United States in 1906.

One of those 100 Indian students in 1906 was Rathindranath Tagore, (son of Nobel Laureate poet Rabindranath Tagore). Rathi later described his first trip to America, bound for Berkeley:[42] He had third-class tickets on a passenger ship, where he shared a cabin with twenty-eight people assigned to five tiers of bunks. The ship arrived in San Francisco in the morning of the fateful day of April 18, 1906—the day the great earthquake and fire devastated the city. From the ship dock the young student could see the charred remains of a few skyscrapers and the thick black clouds of smoke, his first glimpse of the United States. His letter of introduction for a Berkeley professor was now useless, as the Berkeley campus had been destroyed.

Hearing that there was a good agricultural college at the University of Illinois, Rathi and a traveling companion headed for Chicago, instead, thinking the university could not be that far from the city. Of course, school was not nearby, and he had to take another train from Chicago to Champaign, not knowing how he would manage to get to campus from the station. Deciding it would be a good idea to contact the school in advance of his arrival, he sent a telegraph to the Secretary of the YMCA, Mr. J. H. Miner, asking whether he would arrange to receive Rathi at the Champaign station. To Rathi's surprise, however, when he reached Champaign, nobody was there. He managed to get to campus on his own.

Several days later, Rathi met Mr. Miner and discovered the problem. The telegraph had been changed to read "Two students from Indiana." The lady at the Chicago telegraph office had made the correction herself, doubting there would actually be students coming all the way from India. Indiana being the neighboring state, Mr. Miner did not bother to receive the two students. This and other experiences prompted Rathi to become active in improving services for incoming foreign students at the University of Illinois. He was instrumental in founding its first Cosmopolitan Club, which became an important support system for international students, remained in service for many years, and served as a model for the establishment of other Cosmopolitan Clubs across the country.[43]

Indian student enrollment in U.S. colleges increased gradually over the next decade. Of the 235 Indian students reported by the United States government to be in academic residence in 1921–1922, the University of Illinois and the University of California enrolled the greatest numbers, thirty-one and twenty-six, respectively. By field of study, Indian students were enrolled primarily in engineering and liberal arts programs, followed by agriculture and commerce.

Cosmopolitan Clubs

Prior to 1903, foreign students in American colleges and universities were not provided with a formal organization specifically designed to address their unique needs. Despite the evangelical and humanitarian efforts of churches, the Boards of Mission, the Young Men's Christian Association, and the Young Woman's Christian Association, foreign students often had to confront issues pertaining to social adjustment or other concerns without help. The first organized attempt at forming reliable associations capable of attending to the special needs of foreign students was the founding of the Association of Cosmopolitan Clubs of America (ACCA). These informal enclaves, more commonly referred to as "Cosmopolitan Clubs" or "International Clubs," were initiated and developed by the students themselves. For many years these clubs provided a social forum and a support system for nonnationals on campuses across the country.[44]

The first Cosmopolitan Club was formed at the University of Wisconsin. Louis Lochner, a student from Illinois who later became the first president of the Association of Cosmopolitan Clubs, spoke of the initial meeting:

> In March, 1903, sixteen foreign students of the University of Wisconsin, representing eleven different nationalities, met in the modest little apartment of Karl Kawakami, a Japanese student. An international club was to be organized in which all the foreigners of the university, rich or poor, were to meet on an equal basis of mutual friendship and brotherhood. No similar organization at any other university furnished them a precedent. The action of these sixteen men was original, unsolicited, and unprecedented.[45]

Cosmopolitan Clubs spread quickly to campuses around the country. Cornell began a club in 1904; Michigan, in 1905; and Purdue (Ohio) and Louisiana, in 1907. "Above All Nations Humanity" was the anthem heard as the University of Illinois at Urbana–Champaign began its Cosmopolitan Club in 1907. Among its founding members were a newly arrived Russian student, and as mentioned earlier, the son of a world-renowned Indian poet.[46]

In December of 1907 the first convention of the Association of Cosmopolitan Clubs was held in Madison, Wisconsin. By 1912 there were at least twenty-six clubs nationwide.

The forerunners of the organization had been active somewhat earlier in Europe. The first international student congress convened in Turin in 1889. The association that sprung from that meeting was known as *Corda Fratres* ("Brothers in Heart") International Federation of Students and was similar in scope and mission to the Cosmopolitan Clubs. When the American and European groups learned of each other's existence, they arranged to meet and, by 1911, had worked out terms of affiliation. While each group retained its autonomy, they worked together to create a central committee, comprised of two delegates from each representative country. Under this aegis, the affiliates could compare notes and plan joint projects.[47]

Cosmopolitan Clubs typically sponsored an activity called "national night," which celebrated the foods, music, and dances of various countries. The event might be housed in some campus facility or a local church or community hall, and American students were invited to share and learn about the cultures of a school's foreign students. In a time when air travel was not yet an option and few American youths traveled abroad, local teachers used these opportunities as a powerful teaching tool for global education.

The Cosmopolitan Clubs not only provided hands-on assistance for foreign students, but also pushed for more institutional and governmental support. Club members and supporting associates lobbied for universities to appoint special advisors for foreign students to help them comply with the various federal and institutional regulations. Further, club members were successful in convincing the U.S. Bureau of Education to publish an informational bulletin for the guidance of foreign students who were contemplating study in the United States. In 1915, the Bureau published *Opportunities for Foreign Students in American Universities*, which probably was the first official guide for foreign students. It contained information about the departments and features of American colleges and universities, living conditions, college life, and details of entrance requirements for sixty-two institutions.[48]

Cornell's president at the time, Jacob Gould Schurman, was among those who recognized the value of the services the clubs provided. He commended the student-led organization:

> The organization has met a distinct need and is charged a special and most valuable function. On the one hand it has provided a meeting place for foreign and American students. And it has especially promoted acquaintance, intercourse and good-fellowship among foreign students.

Our foreign students in the Cosmopolitan Club have laid before our American students accounts of the contributions which their respective countries have made to the civilization of the world, and compelled from them a respectful and even sympathetic consideration.[49]

The growing influence of the Cosmopolitan Clubs paved the way for other changes in how schools dealt with their foreign students. One exceedingly important development the clubs promoted was the creation of the foreign student adviser (FSA).[50] This fledgling idea varied dramatically from institution to institution. On some campuses, the foreign student adviser was no more than an interested faculty member who made him- or herself available for counsel. At other institutions, formal positions were implemented within the framework of a school's administration. Oberlin College was one of the first to appoint an "official" foreign student adviser, around 1910.

This new concept of providing separate and specialized advisers to address the unique needs of foreign students was described in a 1928 work published by the American Council on Education: "In every institution the foreign student will find an official who is ready to confer with him regarding his special problems," it was asserted. "These may concern his admissions credentials, or the choice of his curriculum, his financial problems, his living conditions, his health, his religious or other quandaries."[51]

As foreign study became more regulated, particularly due to new immigration laws implemented during the early part of the century, the need for foreign student advisers increased. Colleges demanded officials who could oversee the admission of foreign students and ensure that those students adhered to both federal and institutional requirements. Part of the rationale for the creation of the FSA was the fact that foreign students frequently, often unwittingly, put themselves at risk of deportation through their misunderstanding of the regulations.

In the early 1900s, the responsibilities for overseeing foreign students were shared by the Bureau of Citizenship and Immigration Services and the colleges and universities that foreign students attended. Prior to admitting nonnationals, institutions had to petition the bureau for accreditation, and then work with the bureau and make regular reports. Failure to do so put a school at risk of being removed from the bureau's roster of approved institutions. Hence, centralizing knowledgeable FSAs for the purpose of maintaining those mandates was in the interest of students and institutions.

Although the Cosmopolitan Clubs were instrumental in initiating the field of Foreign Student Advising, and had helped to initiate many student support programs and traditions, the ACCA itself did not last. As other agencies and institutional efforts replaced the services originally offered by the Cosmopolitan Clubs, membership faltered and the group gradually dissolved.

The International Houses

In 1909, a young Harry Edmonds, then working for the New York YMCA, had a chance encounter with a Chinese student on the steps of the Columbia University library. "Good morning," Edmonds said casually to the student as they passed. The young Chinese stopped in his tracks, turned to face Edmonds, and said, "I've been in New York three weeks, and you are the first person who has spoken to me."[52]

Edmonds was surprised and unsettled by the young man's remark. Was it possible that foreigners studying in America were truly this isolated? Edmonds soon made it his mission to find out, actively investigating the state of foreign students in New York City. One of his first steps was to author and administer a survey of the estimated 600 foreign students living in the city at the time, to learn firsthand about their experiences, living conditions, and problems.[53] Edmonds discovered not only an almost universal lack of support services for international enrollees, but also a clear and immediate need for institutions to step up to the task of caring for the foreign student populations.

Sunday suppers soon were initiated and brought together foreign students, faculty, and local citizens for friendly exchange (an activity from which the intercollegiate Cosmopolitan Clubs emerged). The Sunday suppers and their increasing community support were a beginning, but Edmonds had much grander plans, which would require substantial funding. He envisioned impressive brick structures in prominent city locations where foreign students from all over the world would reside, dine, enjoy activities together, and have spaces for meetings or study. These "international houses" would provide a central place for community affairs, banquets, fund-raising events, global education, and cross-cultural interaction. They would encourage and promote international tolerance and understanding.

To acquire the necessary funding, Edmonds appealed to the generosity of John D. Rockefeller Jr. the fifth child and only son of

John Davidson Rockefeller Sr., the founder of Standard Oil. Rockefeller Jr. was intrigued by the increasingly popular Sunday suppers and by Edmond's plan for making the city a better host for foreign students and scholars.[54] Edmonds also approached the Cleveland H. Dodge family, which had helped found the Committee on Friendly Relations Among Foreign Students. Both parties were convinced of the need for Edmond's project and agreed to support it.[55]

In 1928, Edmonds witnessed the realization of a vision that had been inspired by a simple "Good morning" in 1909. The first International House opened in 1928 in New York City, at 500 Riverside Drive. The hope "That Brotherhood May Prevail" was boldly inscribed in granite above the main entrance.[56] International House instantly became the most sought-after residence and gathering place for foreign graduate students in the New York area, as well as a venue for all sorts of community and social events.

The success of New York's International House made Edmonds and Rockefeller eager to expand the idea to other areas of the country, and they soon approached the University of California at Berkeley with a similar plan. San Francisco was historically a major point of entry for foreign students, and Berkeley the largest recruiter of international enrollees on the West Coast. It seemed the natural choice for the next venture.

The ambitious project was and immediately controversial and met resistance from the 1920s Berkeley community. First, many people opposed the idea of men and women living together under one roof, as was the practice in New York's International House—coeducational living was hardly an accepted idea at the time. Also, there was mistrust, bias, and occasionally hostility toward "foreigners."

Moreover, the notion that people of color and "whites" would live in an integrated setting was unacceptable to some people.[57] More than 800 community members assembled at a community meeting to protest racial integration in the proposed facility. The most notable and influential speaker at that meeting turned out to be an intriguing woman named Delilah Beasley. Her persuasive remarks at the podium in support of Edmond's plan proved to be a powerful force in countering the racist views of some of those in attendance.

Born in Cincinnati, Ohio, of African descent, Beasley's career began at the age of twelve when she became a correspondent for the Cleveland *Gazette*. Three years later, she published her first column in the Sunday Cincinnati *Enquirer*. She moved to northern California in

1910, attending lectures and researching at Berkeley and writing essays for presentations at local churches.

Beasley also worked for the Oakland *Tribune*, where she wrote a Sunday column called "Activities among Negroes." She spent nine years studying black life in the golden bear state, and in 1919, documented her findings in *The Negro Trail-Blazers of California*. Beasley's influence through the news print media was vast, and her fame was widespread. She had a powerful impact on the actions of the press, and the media image of black America. Chiefly through her effort, the white-owned and -run press discontinued using the words "darkie" and "nigger" as terms for people of color, and began to capitalize the "N" in Negro.[58] Respected by both the white and black communities for her journalistic accomplishments and efforts toward racial understanding, Beasley's outspoken validation of Harry Edmond's proposal was important in helping to establish public acceptance of the project.

Bolstered by increasing public support, Edmonds and his $1.8 million of Rockefeller funding headed to Berkeley to establish a site. He purposefully chose a location on Piedmont Avenue, the traditional home of Berkeley's sorority and fraternity houses, which at that time summarily excluded any foreigners or people of color. By proposing the site, Edmonds sought to strike bigotry and exclusivity "right hard in the nose."[59]

George William Kelham, a prominent local architect, was selected to draw the plans for the structure. The choice was at least partly symbolic, as Kelham had been chief architect for the 1915 Panama Pacific Exposition, an event important in celebrating the art and architecture of many nations.

The original idea was to keep the facility indigenous (owned by International House), like the New York model, to spare the University of California controversy and to allow the students freedom of movement that might not be possible if the property were under university ownership. But, because of tax issues better addressed through the university, the new International House was set up as a separate corporation whose board of directors was interrelated with the institution.[60] International House Berkeley officially opened on August 18, 1930. Offering 338 rooms for men and 115 for women, it was the single largest student housing complex in the Bay Area and the first coeducational residence west of the Mississippi. Just like its predecessor in New York, it quickly attracted students from around the globe.

Leon Litwack, a Berkeley history professor and alumnus, described the campus the significance of the International House:

> The International House was in its very creation a radical idea. It was the first interracial living center west of New York. The International House was established with the intent to give students "the full opportunity for frank discussion on terms of equality, the importance of which it is difficult to overvalue," in the words of John D. Rockefeller, Jr.[61]

The first Director for International House Berkeley was, in a sense, hired by mistake. Among the early concerns about the facility was the possibility of attracting overzealous Christian evangelists, perhaps resulting in uncomfortable situations for the Buddhists, Hindus, Muslims, Jews, and other non-Christians residing at the house. Berkeley's President, William Wallace Campbell, offered Allen Blaisdell the first director's job, for an annual salary of $5,000. A respected educator, Blaisdell had developed a passion for cross-cultural awareness during a teaching assignment in Japan, and was an outspoken proponent of international education. Blaisdell accepted the position and agreed upon a starting date of fall 1928. Soon after Blaisdell agreed to take the job, however, it was discovered that he was not only an educator, but also an ordained minister. President Campbell later stated that, had he known, Blaisdell would never have been given the job, due to potential objections by those who would be concerned that Blaisdell was biased against non-Christians. Nevertheless, he remained in the position and proved to be an effective administrator. But, he was under orders to "keep it [his religion] quiet" and avoid openly revealing his ties to the Christian ministry.[62]

Aside from intruding evangelists, there was widespread concern in the community about other perceived threats to the academic sanctity of the new facility resulting from coeducational, multinational habitation at Berkeley's International House. Many detractors feared that students would smoke, that the ideals of Communism might be spread, and that racial intermarriage would occur.

Colorful stories emanated from both the New York and Berkeley International Houses. In fact, one of Blaisdell's first duties as director of the Berkeley house was to rescue a Japanese student who wanted to marry an American. The Japanese girl had been brought to Berkeley by her older male sponsor and settled at International House, where she promptly met and fell in love with the son of a minister from Oregon.

It turned out, however, that her Japanese sponsor had secretly intended to marry the girl himself after she graduated. When the sponsor received news of the romance, he was furious and warned that he was en route to Berkeley to find his young American rival and "gun him down." Fearing for both of the young students, Blaisdell arranged an escape. Hiding the girl in the back of his car, Blaisdell drove from International House to a safe place, and then arranged her immediate passage to Paris. Disaster averted, the girl eventually returned to Berkeley and married the young man from Oregon.[63]

By all accounts, the first two International Houses were successful in their mission to provide a centralized "home" for foreign students and international activities. Following the triumphs of these ventures, Harry Edmonds and John D. Rockefeller Jr. collaborated in the establishment of a third international house, this time in Chicago, which opened in 1932. The International Houses in New York, Berkeley, and Chicago served as models for similar ventures on campuses around the country and the world. For much of the United States, such facilities continue to serve as centers of international education and cultural exchange that affect not only the host institution, but the local and regional communities.

The Committee on Friendly Relations Among Foreign Students

Another organization that significantly contributed to the development of foreign student services was the Committee on Friendly Relations Among Foreign Students, organized in 1911.[64]

The committee was founded under the leadership of the Christian statesman and student organizer Dr. John R. Mott, who later served as General Secretary of the International Committee of Young Men's Christian Associations, Chairman of the World's Student Christian Federation, and Chairman of the International Missionary Council. Other founding members included Cleveland H. Dodge, Andrew Carnegie, George W. Perkins, John W. Foster, Andrew D. White, William Sloane, and Gilbert Beaver.

The Committee on Friendly Relations functioned in various capacities on behalf of foreign students. Its staff members routinely visited institutions where foreign students were enrolled, to consult with the institutions' foreign student advisers, representatives of student bodies, and the foreign students themselves.

The menu of services the Committee on Friendly Relations provided for students was listed in one of the committee's early handbooks, *The Unofficial Ambassador*:

1. Advice and assistance in presentation of scholarship applications.
2. Information regarding the best schools for particular courses of study.
3. Arrangements for them to be met at the boat or train upon arrival and assistance with immigration, customs, and travel bureaus.
4. Furnishing letters of introduction to friends at their destination.
5. Arranging temporary loans in emergencies.
6. Placement as counselors in summer camps.
7. Assisting them to attend summer student conferences.
8. Arranging introductions and hospitality in American homes and occasions to speak before community organizations, especially churches and the YMCA's and the YWCA's.
9. Interesting individual North American students in individual foreign students. This is done through the YMCA's and YWCA's on the campuses.[65]

One of the committee's chief concerns was that "the incoming student have a feeling of confidence and a sense of security" as he or she became acclimated to the new environment.[66] The so-called "Port of Entry Services" arranged for incoming foreign students to be met at ports or airfields in New York City and elsewhere, where they were advised about housing, ground transportation, and in general were provided with a basic orientation to American life.

The committee also initiated a foreign student census, which was begun in 1915 and continued annually for several years. In addition, the organization took on a variety of specific activities, such as coordinating campus and community exchanges, refining service programs, and providing educational campus visitations to help institutions better care for their foreign student populations. The committee's efforts were supplemented by its informational publications, such as *Living in the United States, A Guide for New Visitors* and *Community Resources for Foreign Students, and International Campus*, among others.

Female Foreign Students in the Early 1900s

Information is very limited about the conditions, concerns, and issues surrounding foreign women students in American colleges and universities in the early years of the twentieth century. Programs and scholarships designed for female foreign students were scarce until mid-century. One of the few early initiatives for women students, however, was the Barbour Scholarship program, established at the University of Michigan, through the efforts of Levi Lewis Barbour, a former university regent.

Barbour was compelled to fund the scholarship by the experiences of Kameyo Sadakata. In 1914, this "tiny and timid Japanese schoolgirl" arrived at the University of Michigan to study medicine. There was nothing the 18-year-old wanted more than to become a doctor, but she was soon overwhelmed by the difficulty of adjusting to her new environment.[67] Kameyo sat in her pre-med classes, took notes in Japanese, and then slowly translated them into English; no surprisingly, her resulting grades were poor. "When I think of my unworthiness I feel ashamed, for I have not been doing any good work . . . ," she wrote, "my instructors are patient with me, whose improvement is so slow." Recognizing the difficulties she and other female foreign students faced, Barbour set about establishing a scholarship program that would help finance, house, and support them. With a gift of $100,000, Barbour formalized the program in 1917, and Kameyo was its first recipient. The scholarships were designed for foreign women students who planned to study medicine, public health, or teaching (preferably at the college level) and have remained in force, helping hundreds of aspiring female foreign student earn their degrees. At the time of its inception, it was one of the few scholarship programs in the United States specifically implemented for female foreign students.

Writing in 1925, Katy Boyd George, Administrator for the Committee on Friendly Relations Among Foreign Students, produced one of the few available commentaries relating to foreign women students in America. The report described various elements of the social and religious life of foreign women students. Several interviews with foreign women students also were included.

George concluded that the issues reportedly faced by foreign women students could be organized into a few basic categories: race prejudice; a lack of "earnestness" on the part of American students; overemphasis on the differences among foreign students; lack of unity in the expression of religion; and perceptions of freedom.

Racial prejudice was frequently encountered, according to some of the foreign women students George interviewed, especially when the foreign student was dark-skinned. One East-Indian woman reported being rejected at more than twenty boarding houses because of her complexion. Another student who had encountered a similar situation reported speaking her thoughts when she was refused a lodging place among a group of white women: "I do not mind living with colored girls. Fortunately I do not have the Christian's race prejudice."[68]

Another concern expressed by some of the women in George's interviews was the foreign women students' confusion about the American students' seeming lack of consideration for the academic side of college life. One foreign student complained:

> I find no one in my house [dormitory] who has a sympathy for studies. Some one told frankly, "Oh, I hate them!" We talk about why we come to college, and one girl says "Oh, for anything but study!" I ask her "What for then do you come to college?" and she replies, "Because it is stylish, and Father and Mother wish it." Surely it is good to say so frankly, but how sorry I feel for the poor studies which are so hated![69]

Others expressed resentment of what they referred to as an "overly-kindly" attitude on the part of Americans. It might take the form of over-attention, such as multiple invitations for speaking engagements or social functions, or an excess of deference in general.

Adjustment to new freedoms was another point of concern for some of the female foreign students interviewed. Depending on their country of origin, some suddenly encountered upon arrival in America an environment with choices and options in far greater abundance than they were accustomed to at home. Some only slowly accepted these unfamiliar freedoms, and one returned student offered this advice to foreign women students new to America:

> It is better for a foreign student to be a little too careful than for people to feel that she is bold and overfree. This is especially true of girls' relations with men. Here it is often hard for a girl not living in a private family to know the best customs. It is not at all safe in these matters to follow the example of strangers about her, for their standards may not be the kind by which she would have America judge herself and her countrymen.[70]

The women who were interviewed also described the positive aspects of their American experience, however. Freedom of thought and the opportunity to develop oneself, for example, were listed by some as important benefits of their American education.

Foreign Students and Christianity

American missionary movements in the nineteenth century spread across the globe and contributed to the increasing numbers of foreign students in the United States. Many students came from non-Western societies where American missionaries were active. The missionaries encouraged and helped students from other lands to come to the United States for Christian education, believing that American education was a vital component in the process of supplying future leaders for Christianity. Missionaries of the different denominations tended to recommend students to the educational institutions with which they were affiliated: Methodist missionaries directed the students to educational institutions of the Methodist Church, while Presbyterians, Episcopalians, Catholics, and other denominations did likewise.[71]

In 1925, the Committee for Friendly Relations Among Foreign Students conducted a survey to determine, by country of origin, the approximate percentages of foreign students reporting to be Christian. According to the survey, 30 percent of foreign student enrollments from China were Christian. Approximately 12 percent of Indian students said they were Christians, and approximately 35 percent of foreign enrollees from Japan reported being Christian. The highest percentages of Christians were from the Philippines (88 percent), Korea (90 percent), and Latin America (93 percent).

Responding to the spiritual needs of the many foreign students practicing Christianity in America, foreign student groups helped create Christian support groups to provide venues for worship and communion. The Chinese Students' Christian Association, for example, was organized in 1909 under the leadership of Dr. C. T. Wang, Dr. P. W. Kuo, Dr. David Z. T. Yui, and Dr. W. C. Chen, then students at Yale, Columbia, Harvard, and Michigan, respectively. Maintaining a central office in New York City, the national association's affairs were directed by a Central Executive Board composed of officers of its regional departments.[72]

Local units were formed in various colleges and universities where there were five or more students in harmony with the purpose of the Association. In 1949, the association had approximately 1,200 members participating in 25 local chapters.

The association was affiliated with the Committee on Friendly Relations Among Foreign Students of the International Committee of the YMCAs of North America, the National Committee of the YMCA in China, the Chinese Students' Christian Union of Great Britain, and the Chinese YMCA in France. The association's purpose was defined in a 1931 publication:

1. To cultivate among the members and friends the Christ-like spirit and to apply his teachings in their daily lives and activities.
2. To build strong moral character and to foster the spirit of self-sacrifice and self-development by rendering service to members.
3. To unite those students in cooperative efforts for promoting programs and activities of the Association on various campuses.
4. To study and understand American society; and to interpret Chinese culture and civilization to America through writing, speaking, dramatics and friendly contacts.[73]

The Association provided practical services to Chinese students, particularly new arrivals, who were looking for housing, employment, or other assistance to help them settle in to the new environment. Combined with the organizational and institutional services that were already in place (on most major campuses at least), foreign students now had places to go for help when they hit roadblocks on their academic journey.

The Shaping of Governmental and Institutional Policy

Prior to World War I, foreign students comprised a very small percentage of students on American college campuses. Records vary as to the exact number, but the population remained below five thousand until around 1912. A formal and accurate census of foreign students nationwide was not produced until the mid-twentieth century.

According to figures provided in the Institute of International Education's publication, *Education for One World*, there were 4,856 nonnational students enrolled in American higher education institutions in 1911. That figure increased to 8,357 ten years later, in 1921. It was during this period of growth that both educators and legislators began to anticipate the need for, and take steps toward creating, expanded services and new regulations involving foreign students and education exchange.[74]

The United States Office of Education from its inception in 1867 had been interested in studying and reporting developments in foreign education. In the period from 1889 to 1905, intense American interest in education abroad led to the establishment of the Division of Foreign School Systems, later changed to the Division of Comparative Education, and then to the Division of International Educational Relations.[75]

Though the Office of Education had no official power to direct the action of bodies responsible for education on the state or local level, many of its activities influenced educational relations with other countries. While most of the exchange efforts carried on in the United States prior to World War II were privately sponsored, in at least two instances the government directly involved itself in international education relations. These were regarded later as important precedents for making intercultural relations an integral and instrumental part of the country's foreign policy objectives.

Some of those objectives were defined during the wake of the Boxer Rebellion in 1900, as previously discussed. The U.S. government also was influential in initiating educational exchanges with the Hispanic-American republics. Attempts at cross-cultural cooperation with the region were made as early as 1849, but proved unsuccessful. It was not until 1890 that the situation began to change, when the International Union of American Republics was established for the maintenance of peace and improvement of commercial relations among the countries of Latin America and the United States. By 1906, the union's successor, the International Bureau of the American Republics, had the added responsibilities of collecting and distributing information regarding education. This bureau later developed (1910) into the Pan-American Union, which, among other things, aided in the strengthening of cultural and intellectual ties among the member nations. This, in turn, led to increased numbers of Central and South American students attending U.S. colleges.

The notion of developing America's international education exchange increased in importance over the next several years, bringing more attention to the tasks of recruiting and attending to both foreign students in the United States and American students traveling abroad. In the words of Isaac L. Kandel, author of *United States Activities in International Cultural Relations*:

> Both groups of students—foreign and American—would need counsel and advice in the selection of the institutions best adapted to their needs as well as other pertinent information relevant to travel and study

abroad; and . . . in the interests of international relations and the promotion of good will the flow of students in both directions needed the stimulus of financial assistance in the form of scholarships and fellowships or, in the case of teachers, of the creation of visiting professorships or other aids.[76]

Although government-led initiatives fostered educational and cultural ties among the nations of the Americas, the task of furthering this cause was handled largely by nongovernmental enterprises. By 1925, more than 115 private organizations had been established in the United States that were concerned in some way with global education exchange.

The Institute of International Education

In the spring of 1917, when most Americans were focused on World War I, a Manhattan educator named Stephen Duggan was thinking about the peace that would follow it.[77] Duggan was a Professor of Government at the College of the City of New York and a member of a subcommittee on international educational relations of the recently formed American Council on Education. As a result of his experiences with global cultures, he believed that the trouble between nations had much to do with misunderstanding one another's cultures. Thousands of well-heeled American youth had studied painting in Paris or philosophy in Heidelberg, but at that time few foreign students had seen and sampled American ideas and attitudes. Duggan envisioned two-way scholarships between U.S. and foreign universities. He took his idea to Nicholas Murray Butler, director of the Carnegie Endowment for International Peace.

At the end of World War I, in the fall of 1918, Duggan, Butler, and Elihu Root sat before a log fire at the Columbia University Club to discuss Duggan's ideas for educational exchange and its implications for peace and the future of international relations. The Armistice, which ended the war, had been signed in a railway coach in the forest of Compiegne only two weeks before.

Elihu Root, Secretary of State in Theodore Roosevelt's cabinet, felt that friction between countries was frequently caused by language barriers and inadequate acquaintance of individual nations with other cultures. "Something must be done on a large scale," he said, "to make the people of various countries more familiar with other languages and viewpoints."[78]

Butler, President of Columbia University and a trustee of the newly founded Carnegie Endowment, agreed with Root. He added that, in general, even intelligent and educated Americans knew little about international relations, and emphasized that our "only hope is for us to become internationally-minded."[79]

Duggan expressed the need for a sort of clearinghouse of information that could stimulate interest in the study of international relations and the increasing flow of globally mobile students. His hope was that an exchange of students might prevent the recurrence of another war.

The fireside talk continued late into the evening. By the end of the meeting, the three men had put together a plan that resulted in the establishment of the Institute of International Education (IIE). Its purpose would be to promote a two-way exchange of students and scholars between the United States and all other areas of the world, in every field of study. Participants would be selected according to ability, and upon completion of a program, the students were expected to return to and serve in their home countries. Scholarships would be set up to help support those who had insufficient funds for education.

The Institute of International Education opened its doors on February 1, 1919, in New York City, with Stephen Duggan as its first director, a post he held for twenty-eight years. Armed with the promise of $60,000 a year from Carnegie, Duggan set to work nurturing alliances between U.S. and European universities, selecting the U.S. students to go abroad on scholarships, finding chaperones for the visiting students, and promoting faculty exchanges.

Duggan's original staff members included Arthur W. Packard, subsequently Director of the Rockefeller Brothers Fund, and Edward R. Murrow, later a renowned radio commentator for and Vice President of the Columbia Broadcasting System. The roster of the trustees and the Advisory Council of the Institute included imminent names such as Nicholas Butler, Henry Morgenthau Sr., John Bassett Moore, Alice D. Miller, John Foster Dulles, Joseph P. Chamberlain, Jane Addams, Henry L. Stimson, Harry Emerson Fosdick, Julius Sachs, Charles Evans Hughes, and Quincy Wright.[80]

It was not the intention of the IIE to take on its exchange goals alone. Instead, it served primarily as a mediating agency. In that capacity IIE could establish and manage scholarships, exchange professorships, and student programs of all sorts.

Recognizing the Institute's timely importance, the Ford Foundation provided a grant that helped expand the Institute of International Education by establishing regional offices in the United States. A further grant of $1 million was awarded by the International Education Board (founded by John D. Rockefeller Jr.) and led to the creation of the International Institute of Teachers College, Columbia University.

The organization enjoyed many accomplishments during its early years of operation. It:

- published the first comprehensive list of scholarship opportunities for foreign students in the United States and for Americans abroad; the first guide books for foreign students; and the first monthly publication in the United States exclusively dealing with international education.
- pioneered sending American professors to universities abroad and bringing foreign professors to the United States.
- assisted in securing a change in the immigration regulations, in 1922, to permit foreign students to enter the United States on special visas.
- was instrumental in the establishment of "student third class" travel arrangements on ships, later to be called "tourist class."
- worked to establish more uniform evaluation of academic credentials from all educational systems.
- established, with the United States Department of State, uniform selection standards and the development of binational selection committees, so that only qualified students would come to the U.S. or be sent abroad.[81]

IIE also administered a periodic search for dissertations relating to various aspects of comparative education, which it published and made available to its members. These publications, together with the institute's *Educational Yearbook*, formed a thorough history of world educational developments between the world wars. Efforts to resolve issues related to international education, such as those brought forth by IIE and other organizations, along with immigration legislation, served to shape America's policies relating to education exchange and foreign students for the balance of the century.

Exclusions and Quotas: Immigration Policy Developments

The Immigration Act of February 14, 1903, transferred the administration of the Bureau of Immigration from the Treasury Department to the newly created Department of Commerce and Labor. As the century progressed, new categories of persons were periodically added to the list of those prohibited from entering the United States. In 1903, epileptics, professional beggars, and anarchists were excluded. Four years later, imbeciles, the feeble-minded, those with tuberculosis, persons with physical or mental defects, and persons under age sixteen without parents were barred from entry.

Naturalization was Congress's next point of concern. The process of bestowing citizenship was assigned to Congress by the Constitution, but for more than a century had actually been carried out by any court of record. An investigation in 1905 reported the absence of uniformity among the nation's more than 5,000 naturalization courts. Congress responded with the Basic Naturalization Act of 1906, which framed a standardized set of rules. This act also proscribed specific naturalization forms, encouraged state and local courts to relinquish their naturalization jurisdiction to federal courts, and renamed the organization the Bureau of Immigration and Naturalization, to indicate its expanded responsibilities. To further standardize the procedures, the new organization collected copies of every naturalization record issued by every naturalization court. To prevent fraud, bureau officials rechecked those records to ensure that each applicant for U.S. citizenship had been legally admitted—a daunting task, even at this early stage. When the Department of Commerce and Labor divided into separate cabinet departments in 1913, the Bureau of Immigration and Naturalization was further divided into the Bureau of Immigration and the Bureau of Naturalization.[82]

During this time, between 1911 and 1920, almost six million people immigrated to the United States—an unprecedented resettlement that raised public concern and prompted a series of quotas acts and other policy changes. In 1917 Congress enacted a literacy requirement for all immigrants, and the Immigration Act of 1917 restricted immigration from Asia. In 1921 the Emergency Quota Act restricted immigration from any given country to 3 percent of the number of people from that country living in the United States in 1910. The Immigration Act of 1924 limited annual European immigration to 2 percent of the number of people from that country living in the United States in 1890. The same year, the Oriental Exclusion Act prohibited most immigration from Asia.

While early legislation decreased opportunities for foreign immigration in general, exceptions were made for students wishing to study temporarily in America. They were not exempted, however, from having to comply with new procedural requirements and documentation. These duties fell not only on the foreign student, but also on the host college. Over the next several years, colleges and universities began to develop internal systems that would ensure both student and institutional adherence to the complex mandates.

4

The World War II Years and Their Aftermath

> In England, sober dons beetled back early to Oxford and Cambridge, which announced they would open as usual for the Michaelmas term next month. Both universities expect subnormal enrollments. Cambridge began to remove the 16th Century glass from its King's College chapel. Oxford gave up some of its buildings for hospitals.
>
> *Time*, September 18, 1939

As war clouds gathered on the horizon in the late summer of 1939, many observers feared for the future well-being of institutions of higher learning and, for that matter, for Western civilization itself. It was clearly apparent that education exchange programs would be early casualties of the conflict to come—the very programs whose brave, high-minded initiatives were intended in part to avert the devastation that now seemed likely to engulf the globe. There was no question about the need to suspend such programs in view of the threats posed to student safety under wartime conditions. Commenting on the situation in Europe, an early 1939 article in *Time* magazine reported:

> Saddest educator was white-haired Dr. Stephen Duggan, director of the Institute of International Education, founded in 1919 by the Carnegie Endowment for International Peace, to promote world good will by international exchange of university students. Dr. Duggan expected the war to play hob with the education of 8,000 U.S. students abroad, 7,500 foreign students in the U.S. Sadly he announced that his Institute had had to cancel the fellowships of 300 U.S. scholars due to go to Europe this fall. As he prepared to send 100 others to Canada, South America and the Far East, Peacemaker Duggan said stoutly: "I look upon this war as an interlude in our work. We intend to continue stronger than before."[1]

One harbinger of things to come was the large influx of academic refugees who poured out of Europe during the 1930s Threatened with being stripped of their traditional prerogatives of academic freedom in teaching and writing, in some cases facing the prospect of compulsory service to socialist or fascist governments, prison, or even execution, large numbers of Europe's leading intellectual lights flocked to the United States and elsewhere in search of a safe haven. This group included, especially, distinguished German professors fleeing the depredations of the Nazi regime that come to power in 1933. In all too many cases over the course of the next half-dozen years or so, those unwilling or unable to migrate to safety faced an uncertain fate.

With the restoration of peace in 1945, internationally minded educators and political leaders turned to the work of rebuilding global connections. The founding of such agencies as the Germany Academic Exchange Service, among many similar organizations, was an example of initial efforts in the 1940s and 1950s to promote educational and cultural exchanges as in the prewar years.[2] At approximately the same time, the United Kingdom began inviting German students to apply for Rhodes scholarships and other sources of support for prestigious academic sojourns in England. Other countries followed suit—doing so, remarkably enough, at a time when many institutions were still rebuilding, literally, what had so recently been leveled by the war.

> Hermann Nickel was four years old when Hitler came to power. Two years after the war ended, at age 18, he was bound for Union College in Schenectady, New York, to study political science. Sponsored by the IIE and the Rotary Club, Hermann was about to be America's first post-Hitler German exchange student. Hermann has had his sights set on America. . . . Though raised under Naziism, Nickel was no Nazi. As he told an interviewer: "The merit is not my own. I owe my understanding to my parents. If I had had Nazi parents, I would have been a Nazi." Hermann had spent most of the war years in school, but after being inducted into an anti-aircraft unit with his entire class, he determined that he was "willing not to fight for Hitler" and deserted. "I can't imagine what it will be like in America" Hermann told his interviewer, "I've never seen a real peace land. And the ideas Americans must have about German youth!"
>
> *Time*, September 18, 1947

World War II, it was broadly agreed, had threatened civilization as had no other conflict before it. The possibility of another such conflagration, one fought by combatants armed with nuclear weaponry, was as nightmarish a scenario as anyone could imagine. Hence, against the background of growing Cold War tensions, discussions about creating a new international organization charged with helping to keep the peace took on urgency.

American support for institutionalized peacekeeping of the sort envisioned by postwar leaders was deemed essential because much of the rebuilding and recovery of economic stability in Europe depended on aid from the United States. The American economy had been stimulated by the war, to an even greater extent than had been true in the First World War, and, of course, America had been spared the physical destruction suffered by most European countries. Militarily, the United States remained dominant—the practical effect of which was to secure America's position as undisputed world leader, not to mention its standing as a nuclear power. More to the point, with its global influence at an all-time high, the U.S. was now prepared to support the establishment of the proposed successor organization to the old, enfeebled League of Nations. American support, as events were to show, made the practical difference.

Even before the end of the war, world leaders had begun planning for the new global agency, one free of what (with the wisdom on hindsight) were judged to have been the fatal flaws of Woodrow Wilson's ill-fated initiative decades earlier. In 1945, some fifty-one nations attended the founding conference in San Francisco of the fledgling United Nations.[3] Among the initiatives forthcoming from the new organization, it was expected, would be an array of cultural exchanges and opportunities for study abroad.

The Birth of the Fulbright Educational Exchange Program

The horrors of World War II had everything to do with the establishment of the Fulbright Program. They inspired a young senator from Arkansas to invoke new ways to help thwart future conflicts through the sharing of knowledge and cultural understanding. After years of interrupted European-American intellectual exchange, U.S. colleges and universities were eager to rebuild foreign educational resources.

J. William Fulbright was born in 1905 in Fayetteville, Arkansas—a beautiful Ozark Mountain town and home to the University of

Arkansas, where his mother Roberta was the outspoken editor of the local paper, the *Northwest Arkansas Times*. He attended the University of Arkansas, where he was an excellent student, and a quarterback for the Razorback football team. At twenty, he attended Oxford on a Rhodes scholarship, and had the opportunity to travel throughout Europe. He gained a deep appreciation for the value of cultural exchange. Fulbright completed his degree and practiced law, then, at age thirty-four, he was chosen to serve as President of the University of Arkansas—the youngest president of a major university in America. His term was cut short, however, when Arkansas' new governor, Homer Adkins, fired him, at least in part, because of his mother's newspaper editorials, which had been openly critical of his administration.[4] President Bill Clinton, who was well-acquainted with Arkansas politics and with Fulbright, recounted the story in his 2004 autobiography:

> In 1942, with nothing better to do, Fulbright filed for the open congressional seat in northwest Arkansas. He won, and in his only term in the House of Representatives, he sponsored the Fulbright Resolution, which presaged the United Nations in its call for American participation in an international organization to preserve peace after the end of World War II. In 1944, Fulbright ran for the U.S. Senate and for a chance to get even. His main opponent was his nemesis, Governor Adkins. Adkins had a flare for making enemies ... besides getting Fulbright fired, he had made the mistake of opposing John McClellan just two years earlier, going so far as to have the tax returns of McClellan's major supporters audited.... McClellan never forgot or forgave a slight. He worked hard to help Fulbright defeat Adkins, and Fulbright did it. They both got even.[5]

In 1945, Senator Fulbright introduced a bill in the United States Congress that called for the use of proceeds from the sale of surplus war property to fund the "promotion of international good will through the exchange of students in the fields of education, culture, and science."[6] It was an inspired piece of legislation. After the war, allied nations owed money to the United States for surplus American-owned aircraft, equipment, and buildings. Senator Fulbright convinced the U.S. government to allow these countries to keep these assets in exchange for contributing to the establishment of a local Fulbright Educational Exchange Program. The general idea of the Fulbright Act was "while avoiding damage to fragile rebuilding economies, to channel funds from the sale of surplus military equipment into scholarships, literally swords into plowshares of the mind."[7] On August 1, 1946,

President Harry S. Truman signed the bill into law, and Congress created the Fulbright Program, which over the next half-century funded the education of hundreds of thousands of Fulbright scholars from the United States and sixty other countries.[8]

> "I just started to thinking one day . . . [how] friendship between China and ourselves was greatly influenced by the Boxer Rebellion. . . . The result was to make them actually feel that we were civilized . . . and interested in their welfare. That kind of feeling is fundamental to not having war every 20 years."
> William Fulbright, December 1945

The initial legislation, however, provided no dollars for bringing foreigners to America. It would take separate legislation to make it a two-way flow. The United States Information and Educational Exchange Act, popularly referred to as the Smith-Mundt Act, established the programming mandate that still serves as the foundation for U.S. overseas information and cultural programs. The act was passed by Congress and signed into law by President Truman in 1948.[9]

Precedents for managing the Fulbright Program and distributing its allocations had been loosely set in earlier exchange efforts, such as the Boxer Indemnities, which brought Chinese students to America; Belgian Relief in the 1920s; and Finnish World War I debt repayments. Additionally, programs were implemented in 1938 when the U.S. Department of State began coordinating overseas educational and cultural relations. These early initiatives helped supply experienced staff members to get the new program underway.[10]

Even today, there is argument over whether the Fulbright Program actually began in China or Burma. The first agreement for student exchange under the program was signed in China, on November 9, 1947. The program's actual first participants, however, were from Burma.[11]

Burma was home to some of the richest natural resources in the world. The middle of the Irrawaddy River basin contained high-grade petroleum deposits located close to the surface, and the hills of the Shan States were rich in lead, tungsten, and other ores. The Tenasserim peninsula contained tin. Tropical forests, which covered half the country, grew teak and other hardwoods in abundance, and the fertile climate made the cleared areas bountiful producers of rice, a major export for the country.

Prior to 1886, Burmese kings had done little to develop or secure their resources. Thus, when the British discovered the untapped riches and recognized their immense economic potential, they moved quickly to secure them for the Crown. Troops invaded Burma, and it was annexed to British-colonized India, effectively placing it under British royal rule. The new Burmese government, with the help of the British, established a system of laws and an administration custom-designed to develop Burma's resources. The British-led leadership also founded Government College, which was the beginning of Burmese higher education.[12]

Burma's national stability was about to be shaken, however. In 1942, Japan invaded the country and drove British and Chinese forces to countries far north. By monsoon season, the Japanese had occupied all but the northern mountains, and Burma remained under Japanese domination for the next two years.

In 1944, the Allies recaptured Burma. But by then its economy and political structure were devastated.[13] A new, independent Burmese government took the form of a parliament, and a new constitution established the state's right to nationalize any sector of the economy. Realizing that the British had exploited the nation's resources, Prime Minister U Nu and his newly formed government expanded Burmese control of the country, taking over the Irrawaddy Flotilla Company's steamer lines and imposing bureaucratic regulations on other areas, in an effort to wrench economic control from England and return it to the Burmese. Unfortunately, the government's plans backfired: Burma's economy continued to decline. In desperation, Nu and his government appealed to Britain and the United States for help.

The appeal was too late, though, and civil war broke out, ultimately involving more than 70 percent of the population. Taking advantage of the chaos, a number of Communist groups took over many of the southern regions, and Burma suffered many more troubled years.[14]

There was not a more unsettled place on earth to begin the Fulbright Program than in Burma. Yet, that is precisely where it was begun. J. Russell Andrus, Second Secretary of the U.S. Embassy in Burma, was asked to guide the program, because of his intimate understanding of the small country, instead of the USIS staff. Andrus had been a Baptist missionary, then a professor of economics at the Baptists' Judson College and later at Rangoon University.

The Fulbright Program was unique in Burma not only because it was implemented there, but also because it included medical and agricultural education, setting the template for the Point Four technical assistance projects to come.[15] Initially, the program arranged for six American students to come to Burma and several Burmese to study in the United States simultaneously, and the first grants were awarded in late 1948. One went to Rangoon University, another to famed "Burma Surgeon" Gordon Seagrave's "Hospital in the Hills," and a third to Sao Saimong, a member of the royal family who knew Russell Andrus before the war and would become a government minister. Because of Dr. Seagrave's affiliation with healthcare education, the first Fulbright scholars to go to the United States were nursing students.

A follow-up project, headed up by Otto and Helen Hunerwadel a year later, involved agricultural education and home economics. The Hunerwadels were pioneers of sorts who ventured into outlying rural areas of Burma to promote and implement the program. Stories of their adventures became local legend. Otto Hunerwadel developed such a colorful and widespread reputation that he was the prototype for the hero of writers William Lederer and Eugene Burdick's controversial novel, *The Ugly American*, published a decade later. Contrary to the misconception of those who have read only the title, the local residents began calling Hunerwadel "ugly" to distinguish their beloved teacher from the broader population of self-important, self-serving, and better-dressed Americans.

By late 1949, 123 Fulbright grants had been awarded in Burma, China, the Philippines, and New Zealand, as well as in Far Eastern countries where education exchanges with the U.S. already existed.[16] These grants included twenty graduate scholarships for Americans to study in China; six grants-in-aid for professorships and fifteen for research fellowships in China; twenty-six travel grants for Chinese students to come to the United States; and four for visiting professorships in Burma, among others.

Other early alliances included the Turkish Fulbright Commission, established by a binational agreement signed between the United States and Turkey, and the Australian Fulbright Program, both in 1949. (The latter treaty was the first that Australia ever signed with the United States.) That same year, Germany, addressed the Fulbright Program just after that country established the Federal Republic.

From its inception, the Fulbright Program fostered bilateral relationships with the fundamental principle of international partnership remaining at the core of its mission. The growth and impact of its programs influenced foreign student enrollments in the United States and the rest of the world for several decades.

The Continuing Story of the Institute of International Education

The Institute of International Education (IIE) was approached by the Department of State's Bureau of Educational and Cultural Affairs in 1947 to administer the new Fulbright Educational Exchange Program on its behalf. The duties would be handled in partnership with the Council for International Exchange of Scholars (CIES), a private organization and division of IIE.[17]

Prior to taking on the Fulbright duties, IIE had accumulated a considerable amount of experience developing and administering foreign exchange programs. Established in the early part of the century, the institute had expanded to become predominant in the field of international education exchange.

In 1932, a young Edward R. Murrow became IIE Director Stephen Duggan's assistant. One of Murrow's first assignments was to identify European scholars who were at risk in their home countries and arrange for them to lecture and teach at U.S. colleges and universities. These efforts became the Emergency Committee in Aid of Displaced German Scholars (later, the Emergency Committee in Aid of Displaced Foreign Scholars), the early forerunner of IIE's Scholar Rescue Fund, and later enlarged to help other "displaced foreign scholars" fleeing Nazi repression throughout Europe. Supported by the Rockefeller and Carnegie Endowment for International Peace, with generous hosting by American campuses, the committee was able to help hundreds of European scholars successfully relocate to America.

Following his boss Stephen Duggan's lead, Murrow joined the American-Russian Institute, founded by John Dewey and other leading intellectuals for the purpose of promoting peace between the two countries. Murrow worked with the Emergency Committee until he began his long career with CBS News.[18] (His work in this arena helped lead to his future showdown with McCarthyism.)

Other IIE activities during that time included a Summer Institute (1934) at Moscow University, which offered courses on Soviet civilization for American students. The following year (1935), the institute created a framework for Latin American student and faculty exchanges through the Convention for the Promotion of Inter-American Cultural Relations; and in 1935 IIE initiated the American-Chinese Student Exchange. By 1943, the IIE's responsibilities had broadened sufficiently to open a regional bureau in Washington, D.C., to serve as a center of information on international educational exchange.

In 1946, after more than a quarter-century as director of the Institute of International Education, founding member Stephen Duggan retired. During his administration the institute had placed more than 3,000 Europeans, Asians, and Latin Americans in U.S. universities, and had sent more than 2,300 U.S. students abroad.

Duggan was succeeded by his son, the scholar Laurence Duggan. A product of Exeter and Harvard (Class of '25), and former chief of the State Department's Division of the American Republics, Laurence Duggan was chosen for the $15,000-a-year job not by his father, but by a special committee whose members included Stephen Duggan's former assistant (now Vice President of CBS), Edward R. Murrow. The elder Duggan attended the committee meetings as secretary, whose duties included notifying the person chosen to take over.

> Last May Duggan, at 75 keen-eyed, white-bearded and talkative, began to look around for a successor. Last week, beaming Stephen Duggan announced that the man had been found. It was his son.
>
> *Time*, November 4, 1946

As the younger Duggan took over the leadership of IIE, student scholarships were still largely a one-way street.[19] In 1946, the institute brought about 1,000 foreign students to the U.S., but sent only 65 U.S. students abroad. The institute was wary about sending American students to European universities until Europe's rebuilding had progressed further. Other options, such as attempts to arrange exchanges with Russia, had failed, and few U.S. students had opted to go to Latin American universities.

Sadly, after serving only two years as director, Laurence Duggan's life came to an untimely and mysterious end:

> In the raw, early darkness of a Christmas-week evening, Manhattan's slushy 45th Street rustled with the shuffling sound and movement of people. Fifth Avenue's traffic brayed and rumbled close by. But the opened window, 16 floors above the din, was just an anonymous rectangle of light—one of thousands held by the city's glowing towers against the black sky. No one in the streets noticed the man who was silhouetted in its frame. No one saw him start his long, tumbling drop to the street. He fell on a heap of dirty snow. Passersby stopped, turned, and saw him then; a thin, black-haired man lying broken and dying. The curious gathered, and with them blue-overcoated policemen. Then an ambulance nosed up.[20]

Laurence Duggan, forty-three, was survived by a wife and four children. Newspapers reported that during his career he had spent fourteen years in the Department of State—nine as head of the Latin American Division, four as adviser on political relations—and two years as president of the Institute of International Education. Police investigators were unable to establish the cause of his death. Duggan's brown tweed overcoat and his briefcase, which had contained a ticket for an airplane trip to Washington the next day, were found in his office. His left shoe was on the floor; he had been wearing only the right one when he fell, and one of the two windows of his office was open.

The date of Duggan's fall was December 20, 1948, only five days after Alger Hiss had been indicted for perjury by a federal grand jury (Hiss was president of the Carnegie Endowment for International Peace and had accompanied Franklin Roosevelt to Yalta to assist with the creation of the United Nations). The police did not reach any definite conclusions, but Republican Congressman Karl E. Mundt did. Mundt announced soon after the news of the death was made public that Laurence Duggan had attended a secret hearing held by the House Un-American Activities Committee early in December, where Russian-born Isaac Don Levine, editor of the anticommunist publication *Plain Talk*, had made a damaging charge: Levine claimed to have overheard ex-Communist and Soviet spy Whittaker Chambers tell former Assistant Secretary of State Adolf Berle that Laurence Duggan had leaked secret U.S. government information to the Communists.

There was speculation that the FBI had pressured Duggan for information about the charge and he had jumped to his death. The FBI publicly admitted that they had questioned Duggan about the Hiss case at Duggan's home just days before his death, but insisted the interview had been routine.

Secretary of State Cordell Hull sprang to Duggan's defense, along with dozens of others. Laurence Duggan's friend and colleague Edward R. Murrow (who, at the time, was serving as chairman of IIE's board of trustees) spoke bitterly about the press coverage: "A dead man's character is being destroyed . . . the headlines might as well have read, 'Spy Takes Life.' "[21]

A half-century later some are still debating what happened. Writing in 2005, author and columnist Ann Coulter contended that "contrary to the image of the Black Night of Fascism under McCarthy leading to mass suicide with bodies constantly falling on the heads of pedestrians in Manhattan, Duggan was the only suicide"—although she added that considering the people he was associating with, he could have been pushed. After Duggan's death, Murrow, "along with the rest of the howling establishment," had summarily denounced the notion that Duggan could have been disloyal to America. "Well, now we know the truth," Coulter concluded. "Decrypted Soviet cables and mountains of documents from Soviet archives prove beyond doubt that Laurence Duggan was one of Stalin's most important spies. 'McCarthyism' didn't kill him; his guilt did."[22]

At the time of Duggan's suicide there was little solid evidence to prove that Duggan had knowing contact with Communist spies. Amid insinuations that he had been friendly with a former State Department official who was a member of a Communist apparatus, and the *New York Daily News* quoted Hiss as having said that the official and Duggan had been close associates. Hiss later denied ever making the statement.

Richard Nixon, then a California congressman, added, "Whittaker Chambers' statement clears Duggan of any implication in the espionage ring." Nevertheless, Laurence Duggan's death remained unexplained. After a week of investigation the New York police department announced the result of its findings, reporting that "Mr. Duggan either accidentally fell or jumped."[23]

UNESCO

By the end of the decade of the 1940s, the Institute of International Education had taken on the Fulbright Program's exchange duties, in cooperation with the United Nations Educational, Scientific and Cultural Organization (UNESCO). In 1948, the first UNESCO Fellows (foreign specialists brought to the United States for study) were placed and counseled by IIE.

The formation of UNESCO had begun approximately six years before. As early as 1942, representatives of the Allied governments of Europe had met in the United Kingdom for the Conference of Allied Ministers of Education (CAME).[24] World War II was far from over, yet those countries already were looking for ways to reconstruct their systems of education once peace was restored. As the project gained momentum, more governments, including that of the United States, elected to join.

Following a proposal introduced by CAME, a United Nations conference for the establishment of an educational and cultural organization was convened in London in November 1945, gathering representatives from forty-four countries. Spurred by France and the

United Kingdom, two countries that had suffered tremendous hardship during the war, the delegates conspired to create an organization that would embody a genuine culture of peace.

At the end of the conference, thirty-seven countries founded the United Nations Educational, Scientific and Cultural Organization (UNESCO). Its constitution was signed in November 1945 and came into force a year later, after ratification by twenty countries: Australia, Brazil, Canada, China, Czechoslovakia, Denmark, Dominican Republic, Egypt, France, Greece, India, Lebanon, Mexico, New Zealand, Norway, Saudi Arabia, South Africa, Turkey, the United Kingdom, and the United States. Paris hosted the first session of the General Conference of UNESCO, which included the participation of representatives from thirty governments entitled to vote.[25]

UNESCO became the only UN body with a mandate to support national capacity-building in higher education. The organization was designed to develop global and regional networks that could assist with a broad range of higher education issues: academic mobility, international exchange, research on education systems and knowledge production, curriculum innovation, leadership roles for women educators, teacher development, and the promotion of quality in higher education. The organization also provided a platform for dialogue on how best to adapt education systems to the new social, cultural, and economic challenges of a globally connected world. UNESCO has played a leading role worldwide in higher education reform.

Smaller Ventures

While IIE, UNESCO, and other large organizations were sponsoring major exchange initiatives, smaller but equally innovative ventures were taking place around the nation. An example of these is the story of Earl Eames and Lloyd Haynes and the Foreign Student Summer Project. One day in 1947, Eames and Haynes, two Massachusetts Institute of Technology (MIT) seniors, began talking about World War II and the devastation they had witnessed in Europe (both men had served in the U.S. Army during the war). By the time the pair had reached their Cambridge subway station, they had come up with a novel idea: Why not help rebuild Europe by bringing top European technical students to the institute, giving them advanced training in their respective specialties, and then returning them home, better equipped to help reconstruct the damage? They decided to formalize a plan.

Finding sponsors was the first problem they faced. Officials at MIT liked the plan and agreed to waive tuition for as many students as the young men could sign up. The rest, however, Eames and Haynes would have to arrange themselves. They began with a trip to Washington, D.C., to get the State Department's blessing, and then proceeded to contact twenty-two foreign embassies, asking each for students.

The next problem was finding funding for the venture. Foundations were contacted, and advertisements were run in local and national newspapers. Eames and Haynes even arranged to be contestants on a popular quiz show called *Break the Bank*, trying (unsuccessfully) to win the $5,000 jackpot. Finally, small gifts began to come in, and enough funds were scraped together to enable the pair to officially approach several students. Twenty-two nations had been invited to send students, and with the exceptions of Russia, Bulgaria, and Hungary, they did. Many countries also contributed money to help with the students' travel expenses to Massachusetts. During the first year, 1948, sixty-two students participated in the program, and the next year, seventy-eight.[26]

By 1949 the Foreign Student Summer Project was an established and successful component of MIT's international effort. The project two ex-GIs had imagined on the way to their subway stop had become a reality, complete with success stories. One West German student who had participated in the project went on to build Bavaria's first electronic computer. Others included a Norwegian who discovered a new method of making gelatin out of seaweed, and a Finn who became an architect and editor of Finland's leading architectural magazine.

The 1948–1949 Foreign Student Census

Fueling projects such as those initiated by MIT was the unprecedented growth of foreign student enrollments in America that followed World War II. Accompanying the escalation in numbers were shifts in proportional representations of students from the various world regions. These shifts were generated chiefly by postwar recovery problems, the complication of monetary exchange barriers, and a general deterioration of relations with "Iron Curtain" countries, as the Cold War deepened.

In addition, reeducation programs begun by the Department of the Army and the Department of State were adding greatly to the flow of students to the United States from former occupied territories: Austria,

Japan, Germany, and the Ryukyuan Islands.[27] Germany, for example, moved from seventh to third in the ranking of represented nations, and Japan moved from twenty-second to tenth. With the rapid increase of foreign student numbers, an accurate and periodic census to track population levels and migration patterns was increasingly important.

In the early years (between 1919 and 1942) whenever a census of foreign students in the United States was attempted, the customary list of schools polled was taken from a directory compiled by the Office of Education. Beginning with the first census, in 1919, the Committee on Friendly Relations Among Foreign Students used the directory, sometimes adding schools they thought were appropriate for inclusion. In 1942, the committee reported 8,075 foreign students from 95 countries studying in 600 U.S. institutions of higher education. The figures were compiled from the enrollment numbers that were reported by those registrars who responded to the census request.[28]

Table 4.1 Foreign Student Enrollment in United States Colleges and Universities 1942–1943

	Men	Women	Total
Afghanistan	8	0	8
Africa	3	2	5
Albania	1	2	3
Arabia	1	0	1
Argentina	82	17	99
Armenia	2	0	2
Australia	21	10	31
Austria	215	105	320
Bahama Islands	3	1	4
Belgium	25	23	48
Bermuda	10	4	14
Bolivia	29	3	32
Brazil	128	26	154
British Guiana	7	1	8
British Honduras	1	4	5
British West Indies	16	5	21
Bulgaria	18	8	26
Canada	821	373	1194
Central America	4	0	4
Chile	78	16	94

(Continued)

Table 4.1 (*Continued*)

	Men	Women	Total
China	598	184	782
Colombia	139	17	156
Costa Rica	67	21	88
Cuba	194	73	267
Czechoslovakia	68	41	109
Danzig	2	0	2
Denmark	14	5	19
Dominican Republic	21	3	24
Dutch East Indies	4	2	6
Dutch Guiana	0	1	1
Dutch West Indies	1	0	1
Ecuador	42	9	51
Egypt	19	2	21
El Salvador	21	6	27
England	193	133	326
Estonia	3	0	3
Fiji Islands	0	1	1
Finland	5	4	9
France	94	92	186
French Indo-China	1	0	1
Germany	462	228	690
Greece	36	8	44
Greenland	1	0	1
Guam	1	0	1
Guatemala	44	10	54
Haiti	48	3	51
Honduras	35	5	40
Hong Kong	3	3	6
Hungary	51	29	80
Iceland	40	8	48
India	45	9	54
Iran	8	3	11
Iraq	28	4	32
Ireland	9	3	12
Italy	60	36	96
Jamaica	24	6	30
Japan	33	10	43
Korea	19	7	26
Latvia	13	4	17
Liberia	3	2	5
Liechtenstein	2	0	2
Lithuania	9	11	20

(*Continued*)

Table 4.1 (*Continued*)

	Men	Women	Total
Luxembourg	0	1	1
Manchuria	1	1	2
Mexico	290	68	358
Morocco	2	0	2
Netherlands	51	34	85
Newfoundland	4	3	7
New Zealand	3	0	3
Nicaragua	39	2	41
Nigeria	13	0	13
North Ireland	2	1	3
Norway	20	6	26
Palestine	43	14	57
Panama	151	38	189
Paraguay	9	0	9
Peru	96	13	109
Philippine Islands	62	27	89
Poland	116	61	177
Portugal	2	4	6
Puerto Rico	388	186	574
Rumania	24	6	30
Russia	74	43	117
Scotland	18	8	26
Siberia	2	1	3
Sierra Leone	4	0	4
South Africa	11	5	16
South America	2	6	8
Spain	25	11	36
Sweden	20	8	28
Switzerland	65	34	99
Syria	8	1	9
Thailand	24	1	25
Trinidad	9	0	9
Turkey	140	16	156
Uruguay	28	4	32
Venezuela	114	10	124
Virgin Islands	7	7	14
Yugoslavia	14	10	24
Stateless	35	13	48
TOTALS	5,849	2,226	8,075

Source: Committee of Friendly Relations Among Foreign Students, 1942

In 1947, when the committee and the Institute of International Education first collaborated to conduct a census, the same list of institutions was used. However, this list did not include schools such as hospitals that provided nursing instruction, theological seminaries, and other business and professional schools, such as those operated by private industry, even if these institutions enrolled foreign students. Therefore, in 1948, the committee and IIE agreed that some of these institutions should be added to the list, so as to produce an accurate count of foreign students in the United States. The final list for the 1948–1949 census of foreign students in U.S. higher education included 2,512 institutions.[29]

Donald J. Shank, vice president of IIE when its first collaborative census was published, wrote in 1949 about the state of foreign students in America:

> At a time when personal contacts between the United States and other countries of the world are of special importance, it is significant to know that there are 26,759 foreign students here this year in our academic institutions. Our academic visitors journeyed from 151 countries, colonies, dependencies, protectorates, absorbed states, and states ... they cover the field from architecture to zoology. As a body they are more representative of the peoples of the world than the United Nations, since the peoples of Nepal and Sarawak, for example, have no direct representation in the United Nations General Assembly, yet two students from each of those countries are with us now. Our foreign visitors are of 152 different faiths. ... Slightly more than one-half of the foreign student body are undergraduates ... one out of every five students from abroad is studying engineering ... one out of ten is preparing for a career in some branch of medicine. ... Agriculture attracts only 4 per cent of the foreign students, being outdistanced by business administration, religion, and education.[30]

Most foreign students flocked to America's well-known institutions. The largest numbers studied at Columbia (1,140), the Universities of California (971), and the University of Michigan (818). The Massachusetts Institute of Technology, Harvard, and Columbia had the largest percentages of foreign students representation among the entire student body—7.1 percent, 5.3 percent, and 4.9 percent, respectively. One striking exception existed to this pattern, however: Montezuma Seminary in Montezuma, New Mexico, which had a foreign student body of 309, or 99 percent of its total

Table 4.2 1948–1949 Foreign Student Enrollment: Top Ten Countries of Origin

Canada and Newfoundland	4,197
China	3,914
India	1,493
Mexico	1,344
Cuba	778
Philippines	660
Turkey	555
Norway	541
Columbia	537
Iran	466

Source: Open Doors, 1948–1949

Table 4.3 1948–1949 Foreign Student Enrollment: Top Ten Institutions

Columbia University	1,140
University of California	971
University of Michigan	818
New York University	724
Harvard University	631
University of Minnesota	402
Massachusetts Institute of Technology	382
Syracuse University	363
University of Wisconsin	340
University of Illinois	336

Source: Open Doors, 1948–1949

enrollment, consisting chiefly of Mexicans studying for the priesthood. Because Montezuma was such a special case, IIE omitted it from its college rankings in its annual census. Had it been included, Montezuma Seminary would have outranked such colleges as the University of Chicago, Purdue, and the University of California at Los Angeles.[31]

According to IIE's 1949 Open Doors report, of the more than 27,000 foreign students enrolled, more than half were hosted by six states: New York (5,000, or 18.7 percent), California (3,098, or 11.6 percent), Massachusetts (2,019, or 7.5 percent), Michigan (1,921, or

7.2 percent), Illinois (1,514, or 5.7 percent), and Pennsylvania (1,064, or 4.0 percent). Regarding field of study, engineering took the lead, followed by medicine and the social sciences.[32]

In the late 1940s, the cost of room, board, and tuition at an American college or university usually fell somewhere between $1,500 and $2,000 per year—expensive for anyone at the time but particularly so for a foreign student who had the added expense of travel. About two-thirds of foreign students were dependent on financial assistance of one form or another, either from their home country, or from an individual, school, foundation, or government agency of the United States. Home-government assistance for foreign students usually took the form of all-expense study grants for students deemed as having exceptional promise. Turkey, India, and the Philippines, in particular, invested in grants for aspiring scholars, sending more than 2,000 students to the United States in 1948.

Financial assistance from the United States manifested itself during the postwar years in an unusual variety of ways, including discounted tuition, housing supplied by a fraternity or other group, or grants from local Chambers of Commerce or Rotary clubs, to name just a few. Americans generously supported student education exchange during the years of postwar recovery. In spite of early Cold War tensions, the general attitude across the nation reflected optimism and a universally revitalized reach for the American dream. Extending the dream to foreign students seeking a U.S. education was an effort that many Americans were happy to encourage.

The following figures are taken from the 1949 issue of *Unofficial Ambassadors*, a publication of the Committee for Friendly Relations, and show the overall growth of the foreign student population from 1930 to 1949. The totals reflect early foreign student census figures compiled by the Committee for Friendly Relations, as well as later counts obtained in collaboration with the Institute of International Education.[33]

1930–1931	9,643
1936–1937	7,036
1939–1940	6,154
1944–1945	6,954
1946–1947	14,942
1947–1948	17,218
1948–1949	25,464
1949–1950	29,813

The National Association of Foreign Student Advisers

The practice of advising America's foreign students began as a local-campus phenomenon. During the postwar years, few campuses had offices or officials charged with the full-time job of assisting and advising an institution's foreign students. Often, the task was undertaken by an assigned person within some administrative department. Sometimes, an interested faculty member took it upon her- or himself to advise the institution's foreign students about regulations, policies, or other concerns.[34] Professor J. Raleigh Nelson is a case in point. Nelson first became interested in foreign students during his undergraduate years at the University of Michigan, in Ann Arbor, where he met and befriended medical students from China. These friends left him with an appreciation of the unique challenges faced by foreign students. Years later, after Nelson was appointed to the University of Michigan faculty, he was instrumental in developing a program for teaching English as a foreign language, implemented in 1911. In the same year, the University's College of Engineering authorized the establishment of a committee on Foreign Students, and Nelson was made chairman of the committee and assumed the title of Counselor to Foreign Students for the College of Engineering. In 1933, he was appointed Counselor to Foreign Students for the entire university, and within five years a section of the Michigan Union had been turned into an International Center, with Professor Nelson acting as its director.

The center and its services expanded quickly. Nelson's course in English grew to become two separate agencies for the teaching of English on the campus: the English Language Service, which helped foreign students enroll at the university, and the English Language Institute, which offered a concentrated, two-month course for incoming foreign students.[35]

Similar initiatives were developing on other university campuses. Small advisory offices cropped up everywhere, and the university staff members who worked in them networked among institutions, seeking informal support and creative ideas. Networking was not limited to university personnel, however. Other people working directly or indirectly with foreign students joined the conversations, including U.S. government representatives; International House directors; IIE personnel; Committee on Friendly Relations members; community program members; evangelists; humanitarians; cultural attaches and ambassadors; the occasional senator or governor; AAUW, YMCA, YWCA, and

Rotary International members; the Pan-American Union; the Rockefeller Foundation; the Ford Foundation; and so on. By the mid-1930s, the array of interested parties was so diverse and so vast that it became clear some sort of coordinated effort was necessary to create a formal organization.

In 1936, the IIE and other sources made the decision to form an "Advisory Committee on the Adjustment of Foreign Students in the United States." By 1941, the membership included Edgar J. Fisher of IIE (serving as chairman), John L. Mott from the International House New York, and seven foreign student advisers or interested persons: Rollin S. Attwood, University of Florida; Gladys Bryson, Smith College; Ben M. Charrington, University of Denver; Charles W. Hackett, University of Texas; Charles B. Lipman, University of California; Martin R. P. McGuire, Catholic University; and J. Raleigh Nelson, University of Michigan.[36]

One of the group's first official tasks was to organize a major conference, which took place in Cleveland on April 28–30, 1942, under the auspices of the Institute of International Education in cooperation with the Department of State, the United States Office of Education, and the Office of the Coordinator of Inter-American Affairs (renamed the Office of Inter-American Affairs in 1945). The scope of the conference was broad, covering topics that emphasized issues and problems that resulted from the war and that foreign students now faced, such as difficulties in transportation, selective service regulations, the evacuation of Japanese-American students from the West Coast, and the need for funds and work opportunities for stranded students. The first conference also addressed special problems relating to Latin American students, due to the particular interest in Pan Americanism and the large enrollments from Latin American countries in the United States.

Only a few persons were formally identified as foreign student advisers (FSAs) at the 1942 conference in Cleveland. At the next convocation, held in Chicago in 1947 (and called by IIE and the Department of State), the number of FSAs exceeded thirty. Still, there were barely enough to adequately take care of the ever-increasing postwar enrollments. There were now 10,341 students from 91 countries enrolled in colleges and universities in the United States, with 19 institutions hosting more than 100 nonnationals on their campuses.[37]

The 1947 Chicago conference, also sponsored by IIE and the Department of State, was responsible for creating a Conference Steering Committee to guide future conferences. That year, the attendees also adopted a resolution to establish a Committee on By-Laws for a "National Association." There was discussion about what to name the new association, as well as about issues pertaining to membership qualification and the structure of the organization, all of which were to be voted on at the next conference, in 1948.

At the 1948 Ann Arbor Conference on International Student Exchange, assisted by the W. K. Kellogg Foundation (at the request of the new Steering Committee), the new National Association of Foreign Student Advisers (NAFSA) was born. Now, 43 of the 224 conference attendees had titles that could be interpreted as "foreign student adviser."[38]

By the end of the 1948 meetings, the bylaws had been discussed, modified, and passed, and officers were elected. Professor Clarence Linton, Advisor to Students from Other Lands at Teachers College, Columbia University, was named president. Allen C. Blaisdell, Advisor to Foreign Students and Director of International House, Berkeley, became vice president. Other officers included Harry H. Pierson, Director of Programs for the Institute of International Education, as secretary, and Joe W. Neal, Adviser, Foreign Students' Advisory Office, University of Texas, as treasurer. Twenty-four others were elected to the board of directors, including Laurence Duggan, President of the Institute of International Education, and J. Benjamin Schmoker, general secretary of the Committee on Friendly Relations Among Foreign Students. Under their direction, all the groups previously mentioned—foreign student advisers, teachers of English, and community groups—were now joined in a combined effort through membership in NAFSA to better serve students from abroad.

Article II of the new association's bylaws stated that:

> The purpose of this association shall be to promote the professional preparation, appointment, and service of foreign student advisers in colleges and universities and in other agencies concerned with student interchange; to serve more effectively the interests and needs of exchange students; to coordinate plans for student interchange through comprehensive voluntary cooperation of all agencies and individuals concerned with exchange students; and in fulfillment of that purpose to initiate, promote, and execute such systematic studies, cooperative experiments, conferences, and such other similar enterprises as may be required to that end.[39]

The quality of the foreign student experience in the United States, both in the classroom and out, has been NAFSA's central concern since its inception. Recognizing that the key people responsible for creating a good experience for foreign enrollees were largely student advisers and English teachers, and that turnover was traditionally high among those groups, NAFSA considered in-service training to be a major focus.

NAFSA's publication program, introduced in 1949, became one of its many and significant contributions to the field.[40] Established with Katherine Bang as its first editor, the widely circulated *NAFSA Newsletter* provided the first vehicle for the dissemination of written information among professionals who advised and otherwise served foreign students.

Four years after NAFSA's founding, its first grant for travel abroad was established with the American Friends of the Middle East, and has continued on an annual basis ever since. Other NAFSA services and contributions have included its government liaison activities, the development of relationships with educational and cultural officers of foreign embassies in this country, its cooperative relationships with other organizations interested in international educational exchange, its development of a code of ethics for practitioners in the field, and the encouragement and reporting of relevant research.

Burden of Proof: Evaluating Credentials

One of the most complex challenges faced during the postwar era by those who served foreign students was determining whether nonnational applicants qualified for admission to American institutions of higher learning. In common with the new army of foreign student advisers and others who assisted international students, admissions personnel faced the task of determining the eligibility of individual applicants. This challenge, in turn, required the development of standards for evaluating students' credentials.

The problem was that foreign students possessed a broad array of types of academic credentials that U.S. college admissions officers were expected to evaluate correctly, determining not only whether a student's accomplishments were sufficient to qualify for enrollment, but also at what level the student should enter the system. This was a frustrating job. Credentials differed from country to country and from region to region. To further complicate matters, it was not uncommon for grades or records to be exaggerated or even falsified.

During the late 1940s and early 1950s, American collegiate admissions had access to governmental assistance when they had difficulty with credentials evaluation. If confused by academic equivalency, admissions personnel could call upon the division of International Educational Relations in the United States Office of Education. Since 1918, this office had housed a small staff of specialists to aid and advise in the processing of evaluating any foreign student's credentials. In the division's first year of operation, there was only request for help. In 1924, there were 81 requests for assistance. Only four years later, in 1928, the number increased to 299. By 1948, however, the office handled nearly 3,000 requests for help from college admissions officers.[41]

In addition to the Office (now Department) of Education, other agencies assisted with foreign student credential evaluation. These included the Institute of International Education, the Department of Education of the Methodist Church, the American Association of Collegiate Registrars and Admissions Officers, and the National Association of Foreign Student Advisers.[42]

In postwar America, admission of a foreign student into a college or university involved a series of steps that began with gathering the student's education records and evaluating the records' equivalency to the requirements of the institution to which the student was applying. The process usually entailed interviews of the applicant and a determination of his or her ability to communicate effectively in English. In addition, many U.S. institutions offered advanced-standing credit by examination.

Each student's country of origin and the characteristics of that country's education system had to be taken into account during the admissions process. To complicate things further, American colleges themselves had differing standards.[43] For example, since 1935, Iran had implemented an education system based on six years of elementary school followed by six years of secondary school, divided into two cycles of five years and one year. Each cycle was concluded by a state examination. When considering an Iranian student's qualifications for admission, many American institutions required a certificate indicating the completion of the fifth year, while other U.S. institutions required the certificate identifying the completion of the sixth year. Advisers could not agree on which level of achievement from the Iranian schools actually corresponded to American expectations for a high school education.

Another example of this problem was India. In 1950, India's public educational system consisted of four years of primary school, a four-year middle school, then two years of high school. After the ten-year program, students took either the "school leaving examination" or an Indian university's matriculation examination. After two years of university work, students were required to take the intermediate examination, followed by an additional two years of college to attain a bachelor's degree. In order to evaluate the credentials of Indian students, American admissions officers took various approaches. Some U.S. institutions reportedly required a certificate of completion of the middle school, while others required the certificate of the school leaving examination and still others required the Indian university matriculation examination. A few American institutions asked for documentation of the intermediate examination before admitting Indian students.

In pre-Communist China, meanwhile, education was comprised of six elementary classes, three years of lower middle school, and three years of higher middle school.[44] At the time, official Chinese policy was to disallow the education of their citizens in other countries. After the People's Republic of China took control in 1949, an isolationist policy was enforced and the education systems and requirements were changed. When Chinese students again ventured into American colleges and universities, admissions officers were faced with evaluating mixed sets of academic records—sometimes even for the same student, whose education had spanned the two distinct systems. English-language proficiency was often the chief obstacle for Chinese students. Many American admissions offices recommended that Chinese students have a year of college in their own country before coming to the United States.

Transfers of foreign students between American institutions caused similar frustrations. While all colleges required a statement of grades and credit points earned, institutions varied dramatically in their requirements regarding advance-standing credit, secondary school units, scholastic aptitude tests scores, achievement tests, or personal interests. Student surveys in the postwar era often showed a high degree of dissatisfaction among foreign students who were aggravated by a multiplicity of systems and incongruous requirements during their transfer experiences.

War and Postwar Immigration Policies

Since the enactment of America's first Chinese exclusion policies in the late 1800s, American immigration legislation had primarily singled out Asians, and for the next half-century many were prohibited from entering the United States. Although students coming into the United States temporarily for education were usually exempted from the exclusion policies, the general mood of the country for a long period must have seemed unwelcoming to such students. World War II witnessed a gradual loosening of the prohibitions, however, allowing for increased immigration and thus creating a generally more inviting environment for Asian students as well as others.

In 1943, for example, the Magnuson Act repealed the exclusion of Chinese immigrants, although, even under the new provisions, only a token 100 Chinese immigrants (selected by the U.S. government) were allowed to enter the United States each year.[45] Under the new policy, the Chinese were granted the right to seek citizenship. Also in 1943, a military order declassified Koreans in the United States as aliens, while the exclusion order against Japanese Americans was revoked in 1945. In 1946, the Filipino and Indian Naturalization Act extended naturalization eligibility to persons of race indigenous to India and the Philippines. The War Brides Act (1945) provided for admission to the United States for spouses and children of U.S. armed forces members; and in 1946, the wives and children of Chinese American citizens were allowed to apply as non-quota immigrants. In 1948, the Displaced Persons Act enabled 15,000 Chinese in the United States to change their citizenship status.

Stranded in America

During the first half of the twentieth century, thanks in part to the use of Boxer Rebellion indemnity funds, China had sent increasing numbers of students to the United States for their education. However, larger world events—the defeat of China's Nationalist government, its retreat to Taiwan, and the triumphant establishment of the Maoist Peoples Republic of China in its stead—combined to eliminate any opportunities for another new generation to come to America to study. As China's civil war wound down and Communist forces prepared to march on Beijing, most of the Chinese students enrolled in American colleges and universities suddenly found themselves cut off from any source of support.

By mid-1949, six months after the fall of Peking (Beijing) to Communism, more than 2,000 students reported that their need for financial assistance had grown urgent.[46] Of the 234 students in the Boston area, for example, 97 were declared to be in "acute distress," while at the University of California, at the other end of the country, 60 out of 90 reported themselves experiencing "extreme" financial exigency.

> At Tiananmen Square on October 1, 1949, Chairman Mao Zedong solemnly declared to the world on behalf of the new China, "From now on, the Chinese people have stood up!" But those who stood up discovered quickly that they could stand only on their own soil; the could go nowhere else.
>
> From Quin Ning, *Chinese Students Encounter America.*
> Translated by T. K. Chu

The immediate burden of assisting these stranded students fell on the host colleges and universities. Scholarship funds were made available wherever possible. Tuition payments were postponed and some emergency loans were granted. Local businesses, civic organizations, and religious groups donated both time and money in an effort to help. The Chinese Embassy in Washington also contributed what it could from its rapidly declining funds. These combined efforts were not enough, however, to meet either the students' immediate or ongoing needs.

United States government aid for stranded postwar Chinese began in 1948, when the Department of State (through the Office of Education and the China Institute in America) allocated some $8,000 for distribution to a limited number of students.[47] In the next seven years, that figure increased to almost $8 million and was distributed to the 3,517 Chinese students and the 119 Chinese scholars and professors housed in institutions of higher education throughout America.

A number of sources fed the $8 million funding pool. The first large-scale allocation of funds occurred in April 1949, at the request of the Nationalist Chinese government, when $500,000 was transferred to the Department of State from the aid-to-China program of the Economic Cooperation Administration (which was created to administer the Marshall Plan).[48]

Each funding recipient was required to sign a pledge stating that upon completion of his or her education the person would return to China and make the knowledge and skills acquired in the United States available in the homeland. This condition was dropped, however, when the loss of the mainland to the Communists rendered return to China no longer possible. In July of 1949, a Congressional report stated that:

> The immediate purpose of this legislation is a humanitarian one—to provide urgently needed financial assistance to Chinese students in the

United States. From the standpoint of the long-range foreign policy of the United States, there is an equally compelling reason for assisting these men and women. These students have had an opportunity to observe and experience the democratic way of life. Thus, because of the traditional position of scholarship-leadership in Chinese society, they are in a unique position to exert a profound influence on the future course of their country. There is no question that it is in the interest of the United States to assist these individuals who can play such a vital role in shaping China's future.[49]

The following year an additional four million dollars was made available by Congress, as part of the Foreign Aid Appropriation Act of 1950 (Public Law 327). In June 1950, an additional $6 million was provided under the China Area Aid Act (Public Law 535). These funds were administered by the Division of Exchange of Persons in the State Department. Congressional appropriations were authorized to enable Chinese students and scholars to achieve a meaningful educational objective, after which they could either return to their home country or, with authorization from the Immigration and Naturalization Services, accept employment in the United States to support themselves until such time as a return home was feasible. To qualify for U.S. government aid, Chinese students had to be citizens of China as evidenced by a passport or other identifying document; in financial need as certified by their official campus representatives and references; enrolled in an institution of higher learning approved by the Secretary of State; in good standing and working full time for an academic degree; and residing in the United States before January 1, 1950.[50]

The cost of the funding program was offset somewhat by allowing the Chinese students to earn some of their expenses. Foreign students were normally restricted from working while in the United States, but an amendment to the China Area Aid Act of 1950 authorized the Commissioner of Immigration and Naturalization to promulgate a general regulation permitting Chinese students to take jobs for as long as they remained in America. Of the estimated 1,300 Chinese students who departed the country, the travel expenses of about 930 were paid in whole or part by the United States government.

The greatest number of Chinese students (2,894) was aided by these funds the following year, but numbers steadily declined after that. In another five years, the program was closed and the unused balance reverted to other government programs.[51]

The funding given the Chinese students was motivated by a still-prevailing postwar enthusiasm for educational and cultural exchanges that, many hoped, might help maintain peace among nations. As President Harry Truman soon told American citizens, educational exchanges also were essential if the United States was to combat the misinformation emanating from the nation's adversaries.

5

Foreign Students, McCarthyism, and the Cold War

"This is a struggle, above all else, for the minds of men."

Harry Truman, 1950

Amid the exhilaration generated by the end of World War II, most war-weary Americans looked forward to peace at last and a return to "normalcy." As world events revealed, however, the United States soon found itself drawn into increasingly ominous "Cold War" confrontations with its erstwhile wartime ally, the Soviet Union.

At the Hotel Statler in Washington, President Harry Truman spoke before the annual convention of the American Society of Newspaper Editors in April 1950:

> Communist propaganda is so false, so crude, so blatant, that we wonder how men can be swayed by it. We forget that most of the people to whom it is directed do not have free access to accurate information. We forget that they do not hear our broadcasts or read impartial newspapers. We forget that they do not have a chance to learn the truth by traveling abroad or by talking freely to travelers in their own countries.... All too often the people who are subject to Communist propaganda do not know Americans, or citizens of other free nations, as we really are. They do not know us as farmers and as workers. They do not know us as people having hopes and problems.... They do not even know what we mean when we say "democracy".... This presents one of the greatest tasks facing the free nations today ... to meet false propaganda with truth all around the globe ... We should encourage many more people from other countries to visit us here.... We should find more opportunities for foreign students to study in our schools and universities.... We are already doing some very good work-through the "Voice of America" and the United States information

offices . . . through the exchange of students, through the United Nations and its affiliated organizations. . . . But events have shown, I believe, that we need to do much more. . . .[1]

From a global perspective, it seemed to alarmed observers that the international Communist conspiracy was on the march everywhere, progressively consuming country after hapless country around the world. The litany of victims grew longer and longer: Czechoslovakia, East Germany, Hungary, Yugoslavia, Romania, Bulgaria, Estonia, Latvia, the Ukraine, Poland, China, Cambodia, Albania, Laos, Vietnam, and so on. Closer to home, in the United States, many feared that hard-line Communists, their sympathizers, and other assorted leftist subversives were busy burrowing into the foundations of American institutional life itself—infiltrating and undermining the government, subverting business and industry, corrupting the universities, and even contaminating the ranks of teachers in the lower schools.

The response to this apprehension, spearheaded by Wisconsin Senator Joseph McCarthy, was a campaign to root out all communists, socialists, assorted leftists, and other supposed subversives thought to pose a threat to the nation's security. Many legislators were intimidated by the strength McCarthy's efforts seemed to be gathering, and few were willing to publicly raise an opposing voice. The exception was William Fulbright. Bill Clinton reflected on Fulbright's independent stance in his 2002 autobiography:

> Despite his civil rights stance, Fulbright was far from gutless. He hated sanctimonious demagogues parading as patriots. When Senator Joe McCarthy . . . was terrorizing innocent people with his blanket accusations of Communist ties, he intimidated most politicians into silence, even those who loathed him. Fulbright cast the only vote in the Senate against giving McCarthy's special investigative subcommittee more money . . . back in the early fifties, a period so vulnerable to anti-Communist hysteria, McCarthy was the nine hundred-pound gorilla. Fulbright took him on before the others would.[2]

Colleges and universities made especially tempting targets during the so-called "McCarthy era." "Witch-hunts" were conducted across the nation in search of disloyal professors, administrators and students. Foreign students, scholars, and faculty from countries sympathetic to Communist ideals were scrutinized with special care. Even the signing of a loyalty oath affirming one's unqualified allegiance to the nation might be insufficient to protect a person suspected of subversion. In some instances, security concerns were taken to ridiculous lengths.

One story, for instance, recounts how a group of graduate students at the University of Chicago wanted to have a coffee vending machine installed outside the Physics Department for the convenience of people who worked there late at night. The students started to circulate a petition to the Buildings and Grounds Department, but their colleagues refused to sign: They did not want to be associated with allegedly radical students whose names were already on the document, for fear of being put "on the list" of individuals under suspicion.[3]

Some institutions were less willing to enforce the investigation for "un-American" suspects than others. A contemporary article in *The Harvard Crimson* recalls the institution's legendary protection of its faculty and students during the "Red Scare." Harvard President Lawrence Summers reflected, "I think we all, as members of this community, recognize that one of this community's proudest moments was the way in which it stood up for the rights of its members in the time of the McCarthy period."[4] His claim of heroism notwithstanding, Summers's assessment of Harvard's record as being sterling was not entirely accurate. Closer scrutiny reveals a public façade not quite consistent with reality: Harvard's leaders did resolve to stand firm against government pressure, but at the same time, Harvard privately conducted its own investigations. Several faculty members were relieved of their responsibilities soon thereafter.

The McCarran Legislation

Paranoia helped create an environment conducive to the passage of draconian legislation aimed at eradicating the Communist menace on the home front and countering the dangers posed by the advancing threat. The most comprehensive and far-reaching of these legislative initiatives was the McCarran-Walter Act, Public Law 82-414, otherwise known as the Immigration and Nationality Act (INA) of 1952.

In essence, the new law imposed more rigid restrictions on entry quotas to the United States, codified the scattered provisions of several earlier legislative works (for example, the National Origins Act of 1924), tightened controls over aliens and immigrants, removed racial barriers, and made citizenship available to people of all origins for the first time. It also required screening of aliens to help eliminate suspected security risks. Other statutes addressed questions pertaining to political asylum, deportation, refugee status, intercountry adoptions, replacement of immigration documents, visas, family- and employment-related immigration, foreign student authorizations, travel documentation, lawful permanent residency, employment restrictions, and a host of other provisions.

> "The INS [Immigration and Naturalization Service] really had their act together [in 1960]. I mean, when you arrived in New York they took you over in a room and checked your visa out. I was interviewed for 30 minutes at the U.S. consulate, and I still remember showing them my little X-ray that I had to get for tuberculosis. And then, of course, I had to declare that I never would join a Communist organization. So this is cold war...."
> David Ward, President of The American Council on Education
> (and former foreign student)

Although the INA allowed students to come to the United States as temporary residents, the McCarran legislation increased dramatically the extent of evaluation and documentation required of them. It likewise restricted a foreign student's ability to work while studying in the United States. Many critics voiced their disapproval of some of the INA's most constrictive features, most particularly the limitations placed on global education exchange. The National Research Council of the National Academy of Sciences' American Association for the Advancement of Science (AAAS), as a case in point, was a highly vocal opponent. A statement released by the AAAS in December of 1951 announced:

> The Council of the American Association for the Advancement of Science is profoundly disturbed over the present world conditions which so severely impede the free interchange of knowledge even among friendly nations. Danger to the future of our nation is implicit in such restrictions. The Council recognizes the need for measures which will effectively safeguard our security, but expresses its troubled concern over the manner in which such measures, in particular the McCarran Act, are being administered, to prohibit American citizens from going abroad and citizens of other nations from coming here to interchange knowledge of science which does not affect security.[5]

The year the McCarran bill was enacted (1951), there were 29,813 students from abroad studying in American institutions of higher education, coming from 121 countries and areas of the world. Canada was in the lead, with 4,498 international students, and China took second place, with 3,549. Germany had jumped to third place, with 1,264, displacing fourth-place India, with 1,136. Mexico, with 871 students, and the Philippine Islands, with 839, also were major exporters of students to the United States.[6]

Several factors accounted for shifts in foreign students' proportional representation from the five principal areas of the world: the Western Hemisphere, Europe, Asia and the Near East, the Pacific Islands, and Africa. Postwar recovery problems loomed as a major determinant in many cases, effectively precluding impoverished countries from supporting students who were applying to study abroad in America. Monetary exchange barriers also posed major obstacles, as did growing political tensions between the United States and Soviet satellite countries.

During the 1940s, students from the Western Hemisphere accounted for more than half of all international students living and studying in the United States.[7] By 1950, approximately, the number of South American students had substantially decreased. The number of students from the Western Hemisphere remained around 10,000, but the total number of students from Europe suddenly increased by more than 1,500, making Europe the second-largest contributor of students.

Regarding gender, during the early 1950s the number of female foreign students accounted for roughly one-fourth of the total foreign student population attending U.S. institutions. In 1953, for example, of the 34,000 foreign students enrolled in higher education in the United States, approximately 8,000 were women. Philippine students had the highest ratio of women (44 percent of the total Filipino students). French and German students also included higher percentages of females, while India and Iran had the lowest percentage of females represented. By late 1950s, the ratio of men to women foreign students was closer to three-to-one, with females making up 47 percent of all Philippine students, 39 percent of Jamaican students, 34 percent of French students, and 32 percent of Chinese students.[8]

Taking into account the overall growth of the foreign student population, the changing student migration patterns, the ongoing process of postwar reconstruction, and America's new legislation, hosting foreign students on American campuses during the middle of the twentieth century presented new challenges. M. Brewster Smith, writing for *The Journal of Higher Education* in 1955, remarked:

> Not so long ago, foreign students appeared to most college administrators as a marginal luxury. They brought to the campus a touch of exotic color; they enriched the educational milieu of American students . . . there was no "foreign student problem" then, and little concern with what American education meant . . . to these scattered visitors from abroad. It was the day of the Y secretary and the Cosmopolitan Club. . . . Since the war we have become much more self-conscious about our relations with foreign students. The participation of government in the support of student exchange has made explicit the statement of objectives of national policy. No longer are we content to maintain the campus as a free market place of skills and ideas, under the rule of *caveat emptor*.[9]

Atoms-for-Peace: Deploying Knowledge

In 1953, recently back from a Big Three conference in Bermuda (a meeting among the leaders of the United States, France, and Great Britain), President Dwight Eisenhower offered nuclear materials to an international agency designed to develop peaceful uses of the atom: "The United States pledges before you—and therefore before the world—its determination to help solve the fearful atomic riddle—to find the way by which the miraculous inventiveness of man shall not be dedicated to his death, but consecrated to his life."[10]

This "atoms-for-peace" pledge was aimed at supplying atomic energy information to countries in need, chiefly those with the least industrial and technical development. To help distribute the knowledge, the United States began a number of training programs for foreign students, who could study atomic energy and then go back to their own countries and share the nuclear know-how required to run atomic reactors. By 1955, thirty-one students from nineteen nations were attending a reactor training school in Chicago, and thirty-two students from abroad had signed up for special courses in radioisotope techniques.

In addition to advanced technical training, many other opportunities were provided for foreign students in the United States. The reeducation programs of the Department of the Army and the Department of State, for example, were adding substantially to the flow of students from the Occupied Territories.[11] These programs, designed primarily to teach the ideas of democracy to talented young people from occupied nations, brought hundreds of students to the United States. Germany and Japan, in particular, sent substantial numbers of students to America, as did Austria, which sent 254 students to the United States in 1950, twice the number it had sent the year before.

Many countries were eager to develop modern industry to help build their economies, including Turkey, Iran, Iraq, the Philippines, Indonesia, Israel, India, and Burma. Collectively, these nations sent hundreds of students to study state-of-the-art technical methods in America. Encouraged by President Truman's Point Four program, which designated support for students from underdeveloped regions, even more students came from the Gold Coast (Ghana), Nigeria, Afghanistan, and other nations for education in America.[12]

Another factor facilitated international students' access to nuclear-age technical knowledge: passage of the Fulbright Act made it possible for the U.S. government to establish travel grants.[13] These grants were offered first to students from Australia, Belgium, Burma, Egypt, France, Greece, India, Iran, Italy, Netherlands, New Zealand, Norway, the Philippine Islands, and the United Kingdom. Grants were later extended to Austria, Thailand, and Turkey.

Throughout the 1950s, varying additional factors influenced the migration of foreign students from different regions: for Mexican and Canadian students, it was proximity; for Filipinos, the strong national ties with the United States and the similarity of education systems; for Japanese and Indians, who came from countries with huge populations with more students than institutions to train them, it was the availability of instruction. Some Chinese students had a different reason for studying in America, however: Since the Boxer Rebellion, the Chinese government had encouraged its citizens to obtain education in the United States, but after World War II and China's Communist revolution, many of these internationals remained in the States as political refugees.

The Bamboo Curtain

In 1950 and 1951, the Immigration Service of the Department of Justice (DOJ), acting under legislation relating to national emergencies, issued approximately 150 orders temporarily preventing the departure of certain Chinese from the United States. The actions were both initiated and justified by the presidential proclamation of a state of emergency in late 1950, at a time when few Chinese had a publicly stated desire to return to their homeland.[14] The detained Chinese (mostly students and scholars) were limited to those individuals who had skills that the DOJ determined might be of use to the Chinese Communist regime.

Detention of Chinese nationals by the United States government during the 1950s was surrounded by both confusion and widespread publicity. One of the most dramatic examples occurred in September 1951, when nine Chinese, who were en route to Asia, were removed from the *S.S. President Cleveland* in Honolulu.[15] This act received wide attention by the press throughout the world. While the detentions generated a great deal of negative publicity, however, little media coverage was given to the millions of dollars that had simultaneously been contributed by American taxpayers for Chinese student support.

Reporter Val Chu, a six-year veteran of *Time* magazine's Hong Kong news staff, conducted a month-long survey of the Communist pressures at work on Chinese students in America. The following is a passage from Chu's report:

> At whatever address they happen to be, dozens of Chinese refugee students will get letters from home this week, and each will bring a heaviness of heart and a torment of fear. Since last March, between 10% and 20% of the Chinese students and ex-students in the U.S. have received such letters, for they are a part of Peking's intensive campaign to woo the refugees back. Invariably they come from the students' families, who may not have written for years. They seldom dwell upon domestic trivialities, but upon the glories of the "New China" and the boundless opportunities to be found there. But the most frightening thing about them is not what they say; it is the fact that they are proof that Peking has discovered who and where the refugees are, and is pressuring relatives to bring about their return.[16]

China launched its campaign against the morale of Chinese students in 1954. At the first Geneva Conference, Chou En-lai accused the U.S. of "persecuting" and "detaining" Chinese students (in particular, 124 technicians held by the U.S. during the Korean War) in order to keep new technological skills from benefiting China. (When the U.S. lifted the ban, however, only half of the 124 chose to return home.) China immediately began to register names of the remaining students' families in preparation for its "letters-from-home" offensive. China needed the skills of these refugees, but it also wanted to do away with what might become a dangerous group of well-educated counter-revolutionaries.[17]

As long as the Chinese students remained in America, they lived in a sort of "no man's land." Many had no legal status and were permitted to remain only on an indefinite basis. Some found it extremely difficult make a decent living, and none could make future plans.

During 1954 and 1955, the detention orders were gradually lifted. This reversal of policy was not necessarily good news for all Chinese detainees, as only a handful actually wanted to go back through the "bamboo curtain" to their homeland. In reality many had used the detainment policies as a sort of insurance against Communist pressure. Trying to return home, but being legally restrained from doing so, allowed Chinese in the United States to both appease the Chinese government and to remain in America. At the close of 1955, only thirty-nine of those previously detained exercised their privilege of returning to China.[18]

One student who returned to China reported being devastated by what he saw. Escaping to Hong Kong, he described what could be in store for others thinking of returning home:

> The luxurious *American President* liner carried home our group of youths full of beautiful dreams and boundless enthusiasm. . . . We sang lustily, "Arise, those who don't want to be slaves". . . . But the lusty group was soon told: "Don't consider yourselves returned students who have drunk foreign waters and therefore are special intellectuals. . . . You must realize you underwent a long-term, poisonous education and have deep-rooted, reactionary tendencies. . . . You evaded the revolution by staying abroad. Today, you should atone for your sins". . . . The more I studied and the more brainwashing I had, the more I regretted my return. In the still of the night, I tossed and turned sleeplessly. Sometimes I thought of suicide. But I felt it was not entirely my fault. It was only my naive thinking that did me wrong. I was only one of the many thousands deceived in the greatest confidence game in history.[19]

At various points during this period, the Department of State and the Immigration Service found themselves working at cross-purposes. While the State Department was operating under legislation enacted to assist Chinese students in completing their education in the United States, the Immigration Service was trying to enforce increasingly complex immigration statutes implemented in response to the situation with Korea and the resulting proclamation of emergency.

At the Geneva Conference in 1954, China pushed for a solution to the problem of what to do about the Chinese remaining in the United States. American officials, however, were reluctant to authorize a third party to investigate the status of these Chinese, concerned that their identification could result in retaliatory action by the Chinese government. In addition, the United States would not permit the Chinese Nationalist government's embassy to be undercut by representatives of Communism.

To resolve the problem, the government of India furnished the channel through which Chinese citizens, on their own initiative, could seek assistance if they felt passage back to China had been obstructed.[20] A year later, in 1955, the State Department reported that the Indian Embassy to date had not represented a single Chinese with departure concerns.

The story of these "stranded" Chinese did not end with the revocation of detention policies or with the funding of the education of the remaining majority. The often-perplexing problem of finding satisfactory employment for these Chinese refugees lingered for years and was complicated by the often-daunting task of bringing their families from China to join them. Some of these people shifted to permanent residence status under provisions of the Refugee Relief Act of 1953.[21] Others married Americans, increasing their preferential status as quota immigrants. In either case, regardless of policy, the vast majority of postwar "stranded" Chinese, and their descendants, chose to make America their permanent home.

Internationalizing America's Campuses

As McCarthyism and restrictive immigration legislation eased during the late 1950s, educators and teaching institutions returned to the business of managing and expanding their enrollments, including their foreign student populations. Most educators were preparing for what was predicted to be a several-year period of general enrollment growth. A report published by the Committee on Educational Interchange Policy in 1957 predicted that the college and university population in the United States would double in fifteen years, and that in order to maintain their ratio of foreign students (1.5 percent of the general student population), institutions would need to attempt to double their foreign enrollments during the same period.[22] The tone of the report indicated the author's doubt that this was a reasonable expectation.

One of the main issues discussed in the report was the concern that if the ratio of foreign students fell during that time, "many of the gains made in the foreign student field in the past may be dissipated and the beneficial effects of visitors from abroad on college and community may decline."[23] These fears turned out to be groundless, in fact, because over the next fifteen years, the foreign student population more than tripled.

The report supported its rationale for the expansion of foreign student enrollments with a number of pertinent observations. First, colleges and universities had a fundamental responsibility to foster and extend communication among students and scholars of all nations. Second, the report claimed, U.S. institutions had an obligation to share their educational resources with other countries. Asia, Africa, and Latin America were looking to America to provide specialized training they could not provide themselves due to insufficient personnel. For example, in Indonesia, one national university with a faculty of less than 100 was overwhelmed by an enrollment of nearly 30,000 students.[24] Maintaining a two-way flow of knowledge between countries was as important to the health and vitality of U.S. educational institutions as it was to those in other countries, and the process needed to begin at the student level. The report further suggested that United States institutions should feel obliged to share their educational resources with other countries, and that the typical American campus was well suited to introduce internationals to American culture.

At the same time, friendships and associations with foreign students would broaden the outlook of American students and the American community. Howard Wilson, a writer on campus influences as avenues to international understanding, summarized the claim:

> In a society such as ours, where public opinion is active and leadership of public opinion is crucial, it seems obvious that all college graduates need to be concerned and reasonably informed about foreign policies, about world cultures, about international relations. That is almost a prerequisite for the successful, democratic conduct of international relations. In the typical institution of higher education a relatively small proportion of the graduates have been exposed to any systematic instruction about international relations. . . . This fact makes it extremely important for the individuals and organizations which seek better understanding and orientation on world affairs for all college

students to utilize fully every possible resource of a non-formal character and to relate these non-formal developments to class offerings as supplementary and coordinate.[25]

A survey conducted in the mid-1950s by the Carnegie Endowment for International Peace backed up Wilson's comments. American students from a cross section of fifty institutions were given a choice of twenty factors that might have influenced their outlook on international affairs.[26] According to the survey, students reported that contact with foreign students was less influential than the mass media, their courses and lectures, and the opinions of parents and friends. Nonetheless, surveyed students indicated that contact with foreign students was more influential than contact with other types of foreign visitors, travel outside the United States, military service, participation in student club activities, or contact with faculty members active in world affairs.

In another study conducted by the World University Service in 1955, American students rated contact with foreign students third in importance after the mass media and courses of study in making them aware of international matters.[27] The results of a third study at Pennsylvania State University showed that individual contact with foreign students was regarded as second in importance only to the influence of the mass media, according to the students surveyed.

Clearly, students in American colleges and universities were aware of the significance of internationals as a part of their higher education experience. Educators, administrators, and politicians were convinced by these reports, as well, and took to heart the need for maintaining or increasing the foreign student population.

Issues of Adjustment

During the 1950s, the adjustment needs of foreign students were beginning to be noticed and for the first time, seriously addressed. A 1952 article described what was becoming a familiar issue on campuses across America:

> In a classroom at little Claremont [Calif.] College one morning last week, a professor solemnly stood up before his class, threw his coat over his arm and, pretending to be a waiter, started handing out menus. The professor was not trying to be funny. Nor did his students

laugh, for they were taking up a highly serious matter: how to order an American lunch.[28]

Maintaining and increasing foreign student populations in the 1950s had to do with academic offerings, but it also was a function of students' ability to adjust to the American campus.

Claremont was just one example of a special nationwide orientation program sponsored by the State Department. Alerted by dismayed college officials about the length of time it often took for foreign exchange students to adjust to American culture, the State Department experimented with solutions by selecting a series of centers where students could go for indoctrination before moving on to the colleges and universities where they were enrolled. By 1952, more than 800 students from 52 different countries were taking part in the orientation.

Claremont's thirty-eight international students represented a typical range of problems. For instance, one "homesicku" (homesick) Japanese boy could not eat fried eggs, and an Indian student refused to shower in the nude ("I shall wear my swim suit"). For these and other students, helping them adjust meant going beyond assistance with settling into dorm rooms or English language training. Practical matters needed to be explained.

Working with the program, colleges such as Claremont helped their international students delve deeper into, and become more comfortable with, American life. Students heard talks by labor leaders, presidents of a manufacturing company, and Republican and Democratic national committeemen. They visited prison farms, television stations, and county fairs. Through these events, American colleges went far toward achieving the goal stated by Claremont Dean Emmett Thompson, who said, "In six weeks, we've got to give them a complete course in Americana."[29]

The 1955 Foreign Student Census

The 1955–1956 census was significant insofar as it marked the last year a count of foreign students in the United States was conducted as a cooperative enterprise of the Committee on Friendly Relations Among Foreign Students and the Institute of International Education (IIE). The growth of the foreign student population and the statistical nature of the work, combined with the committee's desire to focus on student services, resulted in the decision to centralize the census-taking process at IIE alone.

By the 1955–1956 academic year, the international student enrollment in U.S. higher education institutions had reached a total of 36,494 students representing 132 nations. Thirty percent of the total came from the Far East, 23 percent from Latin America, 15 percent from Europe, 14 percent from North America (Canada and Bermuda), and 13 percent from the Near and Middle Eastern countries. Twenty nationality groups consisted of 500 or more students, together making up 71 percent of the total foreign student population.[30]

Twenty nationality groups consisted of 500 or more students, together making up 71 percent of the total foreign student population. Of the leading nationality groups, the one showing the most growth was Korean: from 203 students enrolled in 1948–1949, the size of this group increased to a total of 1,815, taking it from eighth to fourth place

Table 5.1 1955–1956 Leading Nationality Groups in Foreign Student Population

Country	Number of Students in U.S.	% of Total Foreign Students
Canada	4,990	13.7
China	2,637	7.2
India	1,818	5.0
Korea	1,815	5.0
Philippine Islands	1,703	4.7
Japan	1,678	4.6
Mexico	1,303	3.6
Colombia	1,172	3.2
Iran	1,011	2.8
Greece	962	2.6
Venezuela	941	2.6
Germany	778	2.1
United Kingdom	745	2.0
Israel	708	1.9
Cuba	690	1.9
Thailand	659	1.8
Iraq	654	1.8
Jordan	581	1.6
France	548	1.5
Jamaica	500	1.4

Source: Open Doors, 1955–1956

in one year. Student enrollments from the Philippines also increased significantly, from 660 in 1948–1949 to 1,703 in 1955–1956. Indian and Japanese enrollments also increased, although Japan's rank fell from fourth to sixth place. The large number of Chinese students in 1955–1956 reflected the situation in China that kept many of these students from returning home.

The only leading nationality group whose enrollment numbers decreased was Colombia—the first sign of reversal since World War II. The other large Latin American group, Mexican students, continued to increase in size.

Among European students in the U.S., Greeks represented the largest group. German students averaged 1,250 students annually between 1950 and 1953, and leveled off to just under 800 students in 1955–1956.

Iranians composed the strongest representation of students from the Middle East. From a 1948 enrollment of 466, their numbers more than doubled to 1,011 in 1955.

In that same year, foreign students were reported in all forty-eight states, the District of Columbia, Alaska, Hawaii, and Puerto Rico. Concentrations of foreign student were found in certain states, with the largest numbers of foreign students, 5,046, enrolled in New York. California was next, with an enrollment of 4,598, followed by Michigan, with 2,687; Massachusetts, with 2,643; Illinois, with 2,299; Pennsylvania, with 1,559; Texas, with 1,341; and Indiana, with 1,173. The District of Columbia, with 1,159 students, and Ohio, with 1,090, complete the top ten list.

The primary institutions hosting foreign students in 1955 included the University of California, with 1,392 students; Columbia University, with 1,224; the University of Michigan, with 1,070; New York University, with 776; and Harvard University, with 729. Also among the top ten were the Massachusetts Institute of Technology, with 658 students; the University of Illinois, with 633; the University of Texas, with 533; Cornell University, with 526; and the University of Minnesota, with 518.[31]

The ratio of men to women among foreign students enrolled in U.S. higher education was approximately three-to-one in 1955. Students from the Near and Middle East (especially Iraq, Jordan, and Iran) and African countries were represented by the highest percentages of male students. Other national groups, however, included a substantial percentage of female students. Women made up 47 percent of Philippine, 39 percent of Jamaican, 34 percent of French, and 32 percent of the Chinese enrollments, respectively.

In spite of the enrollment increases since the war, American educational institutions as a whole were not overcrowded with foreign students, at least by world standards. Of the nearly 3 million students enrolled in American higher education in 1955–1956, only one in seventy (less than 1.5 percent) were foreign, distributed among some 1,630 institutions. Numbers of foreign students in these institutions ranged from one or two to more than 1,000 (in three universities). Only 12 institutions hosted more than 400. For the sake of comparison, in the same year, Oxford University hosted a foreign student population of nearly 13 percent of total enrollments, and the University of Geneva in Switzerland, approximately 14 percent.[32]

The Training of Physicians

A significant portion of foreigners were in the United States for advanced study in the field of medicine. In 1955–1956, a total of 6,033 foreign physicians from 84 countries were training in American hospitals as interns or residents, representing an increase of nearly 1,000 over the previous academic year.[33] Most foreign doctors training in U.S. hospitals came from the Far East, Europe and Latin America, with large numbers also coming from the Near and Middle East and Canada. For many, it was a struggle to accomplish their dream of earning a medical degree from an important and prestigious American institution. One story, reported in 1959, told the story of a young African:

> Nine-year-old Mungai Njoroge had an infected foot. A kindly doctor at the Scottish Presbyterian clinic calmed the boy by showing him colorful test-tubes in the lab. Fascinated, Mungai decided he would become a doctor when he grew up—a next-to-impossible ambition for a Kikuyu tribesman. Njoroge wanted to go to school in America but was denied papers by the Kenyan authorities. So he took a roundabout route, first via Pretoria where he earned a degree at the University of South Africa; then on to London where he peddled cosmetics until he could secure the permits he needed. Borrowing £60 for passage, Njoroge arrived in the U.S. in 1951 with only 3¢. A fellow passenger provided $1.50 for a room at the Y.M.C.A, and the Committee on Friendly Relations Among Foreign Students lent him $70 bus fare to get to California. He was admitted to Stanford University School of Medicine and earned his M.D. in 1957. At 33, Mungai Njoroge was homeward-bound as Kenya's first U.S.-trained African physician.

Within a few years he would raise a 50-bed hospital, the first in Kenya to be operated for Africans by Africans.[34]

Most of the visiting physicians in the mid- and late 1950s were taking advanced training as residents (69 percent were residents, and 31 percent were interns), a pattern that held true for almost all nationality groups. Of the total number of physicians, 13 percent (807) were women. The gender split varied dramatically depending on the country of origin. More than one-half of all the women doctors studying in the U.S. were from the Far East, with the Philippines accounting for 38 percent of all foreign female doctors enrolled.

Although foreign physicians were enrolled for training in forty-four states, the District of Columbia, the Canal Zone and Hawaii, one-quarter of the total was concentrated in the state of New York. Another quarter was training in Ohio, Pennsylvania, New Jersey, Massachusetts, Illinois, Missouri, Maryland, Michigan, and Minnesota, collectively. Eleven hospitals in the U.S. reported more than forty foreign doctors in training on their house staffs.

The largest number of internationals training to be physicians in the mid-1950s were from the Philippines, with a total of 1,065 students enrolled in United States institutions. Canada was next, with 584; followed by Mexico, with 489; Germany, with 364; Turkey, with 320; Italy, with 260; Greece, with 232; Korea, with 216; and India, with 208.[35]

Hospitals in the Unites States with the most foreign physicians tended to be concentrated in the east. Bellevue Hospital Center in New York City hosted the most, with 101), followed by the Mayo Foundation in Rochester, with ninety-seven; Jersey City Hospital, with seventy-five; and Morrisania City Hospital, also in New York City, with forty-eight.

In 1955–1956, the most popular field of study for foreign students in the United States was not medicine, however. Engineering was the leading field, accounting for 22 percent of the total, or 8,089 students, and was the most frequently reported field of study for students from Colombia, Iran, Greece, Venezuela, Jordan, China, India, Israel, Cuba, and Iraq. The second most popular area was the humanities, with a total of 7,585 foreign enrollees (21 percent of the total foreign enrollment), followed by the social sciences, with 15 percent. Students from Canada, Japan, Mexico, the United Kingdom, and France showed a preference for the social sciences, while Korean students were especially concentrated in the social sciences.

More than half of all foreign student enrolled in the U.S. (56.6 percent) were undergraduates, and more than one-third (37.4 percent) were studying at the graduate level.[36]

New Organizations and Initiatives for Exchange

The growth of the foreign student population in the United States during the 1950s encouraged the expansion and diversification of programs within established organizations, such as National Association of Foreign Student Advisers (NAFSA) and IIE. In 1958, for example, the by-then well-established Fulbright Program implemented a new educational exchange agreement between the United States and Spain.[37] After the initial agreement was signed in October, a Commission for Educational Exchange between Spain and the United States was established in Madrid. Funds up to the equivalent of $600,000 in pesetas were initially made available—an amount bolstered two years later by the addition of the equivalent of $1.65 million pesetas. Of the twenty-four government-regulated institutions of higher education in Spain, eight participated in the initial program, including the universities of Madrid, Barcelona, Valencia, Salamanca, Zaragoza, Seville, and Granada. In the years following its inception, the commission succeeded in providing the opportunity for thousands of Spaniards to study in the United States.

An assortment of other new exchange initiatives also during this period, among them a student exchange program started by the Council on Student Travel and a far-reaching student and cultural exchange effort developed by the America-Mideast Educational and Training Services (AMIDEAST).

These new programs were well timed, because with the launch of the Soviet spacecraft Sputnik on October 4, 1957, Americans were forcefully reminded that peace could hinge on mutual understanding, and that tolerance might best be accomplished through educational and cultural exchanges. Only a month after Sputnik's successful orbit of the planet captured worldwide headlines, the U.S. State Department's Office of East-West Contacts approached the Council on Student Travel, asking it to organize and conduct an exchange of youth groups under the government's Cultural Exchange Program with the Soviet Union.

The Council on Student Travel (later renamed the Council on International Educational Exchange [CIEE]) had been established just a few years before as a multinational effort to help American students going abroad and foreign students coming to the United States, by providing safe and affordable travel.[38] By the mid-1950s, the council had grown to include more and diverse programs in the interest of education exchange.

At the State Department's request, the council contacted representatives from the Soviet Union, and working with these representatives, orchestrated a plan for the first formalized educational exchange between the two countries. The resulting collaboration brought twenty Soviets to the United States for a thirty-day stay, and sent forty-four Americans to the Soviet Union for thirty-eight days. The exchange was a remarkable accomplishment, because no sizable group of American students or professors had officially visited the Soviet Union since 1935, and similar groups from the Soviet Union had not come to the United States since 1931. The effort was successful in part because there was renewed interest in student exchange on both sides. The easing of visa requirements by the United States Congress in 1957 and the loosening of travel restrictions by the U.S.S.R. were influential factors.

In 1958, a Cultural Agreement was signed by U.S. Ambassador William S. B. Lacey and Soviet Ambassador Georgi N. Zaroubin.[39] The two governments pledged to encourage student exchange and other programs. Following the final negotiations, on February 21, 1958, the Council on Student Travel and the Soviet Youth Committee announced in New York and in Moscow that the exchange would take place. It seemed a promising beginning for dialogue between the two superpowers. In an interview many years later, Damon Smith, former deputy executive director of the council, commented on the collaboration: "One of the greatest accomplishments of the Council," Smith declared, "was the success in developing exchange programs with the Soviet Union. These programs were never suspended or closed even during the harsh years of the Cold War. . . . "[40]

Also in 1958, the council inaugurated a special orientation program for students coming to the United States as Fulbright Scholars under agreements with various European countries. The program included a nine-day orientation conducted by a special staff on board ships coming to the U.S., as well as a twenty- to thirty-day program after the students reached the country, during which they were temporarily housed with American host families. The first experimental group contained only forty-three students, but the following year, the council enlarged the program to include 136 participants.

In addition to organizing travel and orientations, the Council for Student Travel (with cooperation from NAFSA, the Committee for Friendly Relations Among Foreign Students, and other organizations) was instrumental in initiating its "Farewell Look at America" program, which gave students the opportunity to visit various places in the United States before returning to their home countries.

Also during the late 1950s, while the Council for Student Travel was successfully orchestrating exchanges with the Soviets, AMIDEAST was developing similar opportunities between America and the Middle East. By the early 1950s, the United States' economic and strategic interests in the Middle East and North Africa had grown considerably. Some prominent Americans were concerned about what they viewed as flawed policies in the Middle East—most of which they attributed to misconceptions about the Arab world and Islam. In 1951, twenty-four distinguished American educators, theologians, and writers met to discuss these concerns and began planning an organization to help bridge the gap. Led by renowned columnist Dorothy Thompson, the group founded AMIDEAST for the purpose of improving understanding and cooperation between Americans and the people of the Middle East and North Africa.[41] Headquartered in New York City, the new organization quickly generated widespread interest and attracted hundreds of members nationwide.

AMIDEAST cooperated with exchange organizations such as the National Association for Foreign Student Affairs (NAFSA) to arrange tours of the Middle East for selected college advisers. Its orientation programs on American culture helped acclimate Middle Eastern students to U.S. academic institutions. AMIDEAST also helped create national associations of Middle Eastern students in the United States, whose members assisted in addressing student needs.

In 1953, AMIDEAST expanded its reach by opening student advising centers in Tehran and Baghdad. Within a few years they had opened seven more country offices throughout the region, as far east as Pakistan. These offices served mainly as resource centers for students seeking information and advice on academic programs in the United States.[42] During 1960–1961, more than 10,000 students visited

AMIDEAST country offices, and nearly 2,000 were placed in American universities. AMIDEAST also helped locate scholarship aid for successful applicants, and it initiated a job placement service in 1956 to help Middle Eastern graduates of American institutions enter the work force upon their return home.

The 1959–1960 Foreign Student Census

At the end of the 1950s, 48,486 foreign students were enrolled at 1,712 institutions of higher learning in the United States. The following census data and tables are from IIE's 1959–1960 *Open Doors* annual report. The 2.6 percent increase over the previous year showed a decrease in the rate of growth from preceding years. As in preceding years, Canada sent more students than any other country, representing 12 percent of

Table 5.2 1959–1960 Foreign Student Enrollments by Country of Origin

Canada	5,679
China	4,546
India	3,772
Iran	2,507
Korea	2,474
Japan	2,168
Philippines	1,722
Mexico	1,352
Venezuela	1,126
Greece	1,095
Thailand	1,006
United Kingdom	993
Cuba	935
Jamaica	902
Turkey	835
Israel	807
Germany	720
Colombia	687
Iraq	675
Lebanon	590
France	572
Jordan	556

Source: *Open Doors*, 1959–1960

Table 5.3 1959–1960 Institutions with More than 400 Foreign Students

Institution	Total Enrollment	Total Foreign Students	Percent of Total
University of California	43,478	1,918	4.4
New York University	32,990	1,580	4.8
University of Minnesota	36,288	1,156	3.2
Columbia University	25,865	1,146	4.4
University of Michigan	26,581	1,109	4.2
University of Illinois	27,089	935	3.5
University of Southern California	18,300	864	4.7
University of Wisconsin	26,678	806	3.0
Massachusetts Institute of Technology	6,233	770	12.4
Cornell University	11,184	751	6.7
Harvard University	13,569	738	5.4
Howard University	4,357	591	13.6
Indiana University	23,531	531	2.3
University of Texas	19,508	487	2.5
Purdue University	15,763	486	3.1
University of Washington	20,411	461	2.3
Ohio State University	23,164	431	1.9
Michigan State University	20,549	427	2.1

Source: Open Doors, 1959–1960

the total foreign student enrollment. Chinese students constituted nine percent, followed by India.[43]

By academic level, the majority, 24,100 students, were enrolled as undergraduates; 18,114 were studying in graduate programs; and the rest, 4,266 students, were unclassified. Academic levels varied widely according to world region, however. For example, there were more undergraduate than graduate students from Latin and North America, the Near and Middle East, and Africa. European students were almost equally divided between graduates and undergraduates. Four countries (Cuba, Mexico, Venezuela, and Iran) had more than 75 percent of their students enrolled as undergraduates. The largest percentage of graduate students from any one country was India (67 percent), followed by 62 percent of students from the United Kingdom, 60 percent from the Philippines, 55 percent from France, and 52 percent from Japan. Once again, engineering was the top field of study among foreign students (11,279 enrolled), followed by the humanities and the natural sciences.[44]

In 1959, all fifty states and the District of Columbia reported hosting international students, with California and New York hosting the lion's share (6,457 and 6,069 students, respectively, more than 25 percent of the total foreign students enrolled in American higher education institutions). Other states hosting large numbers of internationals were Michigan, with 3,259; Massachusetts, with 3,136; Illinois, with 2,889; the District of Columbia, with 2,020; and Indiana, with 1,819.

Eighteen institutions reported having more than 400 foreign students in attendance. The University of California, for the third consecutive year, enrolled the largest number of foreign students. Howard University had the largest percentage of foreign students in relation to its total enrollment.[45]

During the 1950s, the number of self-supporting foreign students enrolled in American higher education was decreasing. By 1959, only 38.3 percent (18,557 students) reported they were supporting themselves with their own funds. Private organizations were helping to finance 13,451 foreign students, a 2.3 percent increase over the previous year. United States government grants had increased, too, and were supporting 5.5 percent of the international population.[46]

6

The 1960s and 1970s

"As Oliver Cromwell once said, 'Paint me with all my warts and moles,' and that's the way we want you to see our country."

John F. Kennedy, 1962

At 3:30 on the afternoon of May 10, 1962, President John F. Kennedy spoke before a small gathering of international students on the South Lawn of the White House. He and the first lady, Jackie, were very interested in international students and education exchange. "All of us, at least many of us, have been foreign students," the president told the group. "The Secretary of State, the Under Secretary of State, the head of the Policy Planning were all Rhodes scholars; the Deputy Attorney General and others; the Chairman of the Senate Foreign Relations Committee. My wife was a student in Paris for a year. I studied at the London School of Economics . . ."[1]

The reception for international students was designed to emphasize the administration's commitment to peace and prosperity through global understanding and education. President Kennedy continued:

> I hope that when you go back—and you will be the future leaders of the country—many of you come from countries which have only a handful, as newly independent countries, of educated men and women. You are their investment in the future . . . and I know that you will regard this great effort which has been made, to bring you here and which you have made, not merely as an opportunity to advance your economic interest but to advance the welfare of your country. I hope that you will think well of this country and recognize what we are trying to do. To run a free society is very difficult. Winston Churchill once said, "Democracy is the worst form of government, except for all the other systems."[2]

When Kennedy spoke those words, more than 50,000 foreign students were enrolled in United States institutions of higher education, from 143 countries and in 1,666 institutions.[3] The Mutual Educational and Cultural Exchange Act (Fulbright-Hays Act), signed into law the year before, had consolidated various U.S. international educational and cultural exchanges, including the translation of books and periodicals, and U.S. representation in international fairs and expositions. It also established government operation of cultural and education centers abroad. At the same time, the Bureau of Educational and Cultural Affairs was established in the Department of State under an Assistant Secretary.

> "Students want to come to the U.S.," says Philip H. Coombs, the State Department's Assistant Secretary for Educational and Cultural Affairs. "This is an asset we should be pretty thankful for. We couldn't buy it."
>
> *Time*, September 8, 1961

Foreign student enrollments were escalating from a number of world regions. African enrollments, especially, had increased from a total of approximately 2,000 students in 1960 to more than 2,800 in 1961, a 44-percent increase. The number of Asian students had increased 12 percent from the previous year, and was approaching 20,000.

More than a third of the total foreign student enrollments were from the Far East. Eighteen percent were from Latin America, 15 percent from Near and Middle East; 12 percent from North America, and 5 percent from Africa. By country, 11.4 percent were from Canada, with 6,058 students, which remained the country that sent the most students. China was second, with 5,304; India third, with 4,835; and Iran was fourth, with 2,880.[4]

Kennedy was personally and genuinely concerned with the welfare of the 50,000 foreign students hosted by America's colleges and universities, and he was active in the establishment of a high level of services to meet their needs. In the archives housed at John F. Kennedy Library resides a memorandum written by the president in August 1961 to the Secretary of State:

> I again have been reminded that we do not have an office in the Government which concerns itself solely with the guidance and welfare

of the some 50,000 foreign students in this country. I am sure you agree with me that this is a serious short-coming on our part and one that should be remedied as swiftly as possible. . . . Since you already have been thinking about the problems in the Department, I wonder if we can't now move fast enough to perform at least some services for the students during the 1961–62 academic year. I feel very strongly that we must not allow these students—especially the Africans and Asians—to leave this country disappointed any longer. I do not think we can leave their welfare entirely to uncoordinated private enterprise. Would you let me know what we will be able to do for them during the upcoming academic year.[5]

By academic level, half of all international students enrolled in American higher education at the time were undergraduates, and approximately 40 percent were studying at the graduate level. That ratio fluctuated by world region, however. Well more than half of the students from the Far East were studying at the graduate level, for example—most in engineering and the physical sciences. Institutions in forty-nine states, the District of Columbia, Puerto Rico, and Guam reported hosting foreign students—with the heaviest concentrations in California and New York.[6]

Four years after President and Mrs. Kennedy's reception for foreign students on the South Lawn, President Lyndon Johnson continued the tradition. The 1964 gathering on the White House grounds was attended by 800 foreign students from 71 countries. Following welcoming comments by the first lady (Lady Bird) and a brief commentary by Secretary of State Dean Rusk, the President was then invited to say a few words to the students:

I am glad that Mrs. Johnson asked me to speak. I thought of the story of the man and his wife who were having an argument. The man's neighbor said to him, "I understand that you and Mary had some words." He said, "Yes, I had some, but I didn't get to use mine." I appreciate Lady Bird giving me a chance to use some of my words this afternoon. . . . You have been studying here in America. Many of you will soon take your place in the forward march of your own society. I hope that what you have learned here will help you advance the progress of your own people, for this is no time for men of knowledge and learning to be above the battle, to stand aloof from the fight for a better world. . . . Your education is a solemn trust. It carries the responsibility of a lifetime of service to your own country and to the world. I am glad that you have had this chance to see America, to know its people, to understand its problems as well as its achievements. For, like your own countries, we are an unfinished society.[7]

Two years later President Johnson supported new legislation promoting education exchange. In his Special Message to the Congress Proposing International Education and Health Programs, the president urged the legislators to share his vision:

> To the Congress of the United States:
> Last year the Congress by its action declared: the nation's number one task is to improve the education and health of our people. Today I call upon Congress to add a world dimension to this task. I urge the passage of the International Education and Health Acts of 1966. We would be shortsighted to confine our vision to this nation's shorelines. The same rewards we count at home will flow from sharing in a worldwide effort to rid mankind of the slavery of ignorance and the scourge of disease. . . . We bear a special role in this liberating mission. Our resources will be wasted in defending freedom's frontiers if we neglect the spirit that makes men want to be free. . . . Half a century ago, the philosopher William James declared that mankind must seek "a moral equivalent of war."[8]

Despite the ongoing tradition of presidential support, however, during the early 1960s the United States government itself contributed to a surprisingly small share of the funding for student exchange, providing only partial sponsorship for just 5,000 foreign students. The greater portion of the task of promoting and managing foreign student exchange was left to private enterprise, which Philip H. Coombs (the State Department's Assistant Secretary for Educational and Cultural Affairs) called "the people's branch of foreign relations."[9]

Based on census and other information pertaining to foreign students enrolled in America's colleges and universities at the time, Coombs offered a profile of the "typical" foreign student:

> The foreign student in 1961 is probably a male undergraduate studying engineering (with social sciences favored among Africans). He is far poorer than his often rich predecessors, and he is culturally more remote from U.S. life. He needs more financial help, more guidance, and more understanding than ever. One of his basic psychological problems is an almost invariable loss of self-esteem after arrival; he feels uprooted and hence resentful. He is shocked at the meagerness of his money; U.S. scholarships do not usually cover living expenses or summer vacations as do Europe's. He finds astonishingly diversified colleges with unpredictable standards. He finds rude waiters, Jimmy Hoffa, demanding children, and kind old ladies who ask Africans if they live in trees. He rarely finds anyone who knows the location of Mali, Gabon or Dahomey, or even of their existence.[10]

Government officials and the private sector agreed. More needed to be done to develop programs and services that would ensure foreign students enjoyed a positive higher education experience in America. Over the next several years, many existing organizations added to their international exchange agendas, and new organizations and associations developed.

Innovations and Growth in Exchange Programs

For African students during the 1960s and 1970s, possession of an American college degree was a major, often life-changing, accomplishment. In Kenya for example, education was synonymous with uhuru (freedom), and that country's citizens flooded U.S. institutions with thousands of scholarship applications every year.[11]

There were concerns besides admissions and scholarships, however. Until the 1960s, African students had, for the most part, been left to their own devices when it came to getting to America. The expense was more than many could afford, which led determined young Kenyans to get to the States by any means possible—a fact that sometimes placed these students in difficult or even dangerous situations. The need for travel assistance, which sparked a political battle between the Kennedy and Nixon camps, led to the "African Airlift."

> "When you're dealing with an African student, you may be dealing with a fellow who will be prime minister in five years."
> Philip H. Coombs, 1961

It began when the African-American Students Foundation lined up scholarships to U.S. colleges and universities for 250 students from Kenya and other British-controlled areas of East Africa. Although airlifts had been arranged and funded since 1959, this new project was a considerably larger undertaking than its predecessors.[12] Kenyan labor leader Tom Mboya, who had rounded up the students, now needed money for their transportation. After repeated rejections from the U.S. State Department, Mboya contacted presidential candidate Richard Nixon to seek his influence—but the State Department turned down Nixon, too. The determined Mboya then arranged to meet with John Kennedy, who immediately agreed to help fund the request with Kennedy family money—Sargent Shriver, Kennedy's brother-in-law (and managing director of the family foundation) recommended they put up the entire $100,000.[13]

When word leaked about the contribution, the Nixon camp renewed its interest in the airlift, possibly in response to political pressure. For example, Jackie Robinson, one-time baseball star and important Nixon supporter, called Nixon in Washington, urging his active participation in helping the students. Nixon agreed to try, and presumably through those efforts the State Department reversed its decision and granted the needed $100,000 to the African airlift project. What resulted was rare indeed—a competition between two political factions to fund an educational exchange project:

> Pennsylvania's acidulous Senator Hugh Scott announced that the students would be coming to the U.S. on a State Department grant, and Jackie Robinson happily reported the news in his New York Post column. Then, when Senator Scott learned that the Kennedys, and not the Government, would be picking up the tab, he took to the Senate floor in a boiling rage. . . . "The long arm of the family of the junior Senator from Massachusetts has reached out and attempted to pluck this project away from the U.S. Government," Scott rumbled. . . . Scott's denunciation brought Jack Kennedy to his feet to denounce him for the "most unfair, distorted and malignant attack I have heard in 14 years in politics."[14]

Among others, Arkansas' William Fulbright, then chairman of the Foreign Relations Committee, questioned the State Department's reversal of their original decision not to fund the airlift project, and demanded full disclosure. In the midst of the confusion and accusations, the 250 African students were the big winners, as their passage was thoroughly funded.

Unfortunately, simply getting the African students to America did not solve the monetary problems many encountered upon their arrival. Misinformation, according to some reports, may have led to students' miscalculations of how much money they would need for living expenses in the United States. Coming to America with small scholarships and very limited funds, many of the unwary scholarship recipients found themselves in sudden and immediate need of financial assistance. A typical example was the Kenyan with a $200 scholarship to a Midwestern university who upon arriving found that he owed additional fees totaling $1,000—a sobering realization for someone supporting six children. The airlift initiative had applied a "self-help" theory, which, unfortunately, left some students frantic and begging for money. "Please help me," one wrote to the British embassy, "because I'm beginning to smell."[15]

A program developed by Harvard's then Dean of Admissions, David Henry (1961), brought 250 students, mostly from West Africa, to some 150 U.S. campuses.[16] Coordinated by the African-American Institute, the "Henry Program" included a rigorous selection system, transportation arrangements, four-year scholarships, and living expenses, all paid for by the International Cooperation Administration. A key feature of the program was its orientation for each of the incoming foreign students, which began even before the students arrived in America on the transatlantic steamer. As part of the orientation, students were encouraged to discuss everything from dating to telephone terminology, and even took practice exams to get the feel of U.S. classrooms. Assessing the merits of both the earlier "airlift" project and the Henry Project, Secretary Coombs combined the two approaches through the Institute of International Education in a proposal for the creation and funding of additional exchange programs.

The IIE: New Contributions

During the 1960s, the Institute of International Education (IIE) established overseas offices in Asia, Africa, and Latin America to meet the growing needs in those regions for information about U.S. higher education. Donor-supported educational services likewise were expanded to meet the increasing demand for information on international education and education exchange.[17] Throughout the 1970s, IIE continued to expand its fellowship services in support of human resource development in developing countries. For example, IIE undertook administration of the Venezuelan government's Fundación Gran Mariscal de Ayacucho Scholarship Program, which assisted nearly 4,000 promising young Venezuelans, many from disadvantaged backgrounds, to study in the United States in fields related to national development.

Also during the 1970s, IIE assumed responsibility for a portion of the USIA (United States Information Agency) International Visitor Program, and also began administering the ITT (International Telephone and Telegraph) International Fellowship Program, which for seventeen years was an exemplary model of corporate involvement in international educational exchange.

> **To the Congress of the United States:**
>
> "In the 1970 Fiscal Year, this [Fulbright] program provided 4,638 outstanding scholars and leaders . . . with intensive exchange experiences which linked the United States with 123 other countries and territories. The major part of this report is devoted to a review of . . . activities designed to provide foreign students with broader opportunities to participate in the life of this country . . . the United States now has a large number of students who come from foreign countries—well over 135,000. . . . These students present the United States with an exceptional opportunity. . . . The professional and personal ties which these students form . . . and the insights they gain into our way of life will help shape their future perceptions of America. . . . Public and private programs which enhance the experiences of these potential leaders can do a great deal to build the human foundations for a more peaceful world. I commend this report to the thoughtful attention of the Congress."
>
> Richard Nixon, August 5, 1971

IIE also worked with the White House and USIA in planning the innovative Hubert H. Humphrey Fellowship Program, which brought accomplished professionals from designated countries of Africa, Asia, Latin America, the Caribbean, the Middle East, Europe, and Eurasia to the United States at a midpoint in their careers for a year of study and related professional experiences.[18] The program provided a basis for establishing long-lasting productive partnerships and relationships between citizens of the United States and their professional counterparts in other countries. Operating as a Fulbright exchange activity, the Humphrey Program was funded by the United States Congress through the Bureau of Educational and Cultural Affairs of the United States Department of State.

The J. William Fulbright Scholarship Board (FSB), appointed by the President of the United States to oversee and supervise the educational exchanges of the Fulbright Program, had overall responsibility for the final selection of Humphrey Fellows.[19] The Fellows participated in programs that combined graduate-level academic coursework with professional development activities—not a degree program, but, instead, a program designed to offer broader professional enrichment. Humphrey Fellows were placed at carefully selected American universities, and their needs attended to by a campus coordinator. Humphrey

Fellowships were limited to one academic year, preceded, if appropriate, by a period of English language training.

The South African Education Program

South Africa's racist apartheid system received increasing media attention during the 1960s and 1970s, attention that shed light on, among other horrors, the clear evidence of educational disparities. There were only 1,400 nonwhite South African university graduates in that country in 1970, contrasted with 104,500 white university graduates—in an area where white citizens composed less than 15 percent of the total population.[20]

The real and potential debilitating effects of an undereducated black population on the region's social, political, and economic development were obvious. In 1979, the Institute of International Education established the landmark South Africa Education Program (SAEP) to increase higher educational opportunities for black South Africans.[21] Designed to redress the extreme social inequities imposed by the apartheid regime, SAEP provided disadvantaged students with the knowledge, skills, and professional credentials required to effectively participate in a democratic society.

During the program's first year of operation (1979–1980), six students from South Africa were sent to the U.S. for education. Twenty years later, when the program ended (2001), nearly 1,700 participants had completed their educational programs and returned to South Africa. The primary fields of study included education, business, law, health administration, and engineering.

The IIE originally designed SAEP, but many partners, including public and private affiliates, contributed to the program's implementation. Policy guidance was provided by a National Council, established by IIE and chaired by Harvard University president Derek Bok. The National Council included such distinguished members as Donald Easum, President of the African-American Institute, and William Carmicheal, head of the Ford Foundation's Office for the Middle East and Africa.[22] In 1979, the Educational Opportunities Council (EOC) was established in South Africa with Nobel Peace Laureate Archbishop Desmond Tutu as its founding chairman.

Beginning in 1983, the United States Agency of International Development (USAID) provided an initial $3 million award for additional SAEP scholarships. Other funds came from USAID, the

U.S. International Communication Agency, the United Nations Educational and Training Program for Southern Africa, the Ford Foundation, the Charles Stewart Mott Foundation, the Carnegie Corporation, and others.

The greatest testimony to the success of the SAEP, ironically, was its closing.[23] With the election of Nelson Mandela as President of South Africa and his award of the Nobel Peace Prize, came international recognition that apartheid would be dismantled and that a new and important stage in South Africa's development had begun.

The 1969–1970 Foreign Student Census

During the 1969–1970 academic year, nearly 135,000 foreign students were reported as enrollees by U.S. colleges and universities, setting a new record. As IIE president Kenneth Holland pointed out, record-setting was no longer remarkable. Since the end of World War II, a new record in foreign student enrollments had been set every year.[24] In just fifteen years, since the 1954–1955 census (the first year IIE standardized the census-taking), the foreign student population in the United States had quadrupled.

As in previous years, male foreign students outnumbered female foreign students by about three-to-one. Predictably, the ratios varied from region to region. While European countries sent about twice as many males as females, the ratio of male to female students sent from the Near and Middle East and Africa was about six-to-one.

As in earlier years, the largest portion, or 36 percent, came from the Far East, and the second-largest, 19 percent, was from Latin America. Fourteen percent came from Europe, 11 percent from the Middle East, and 10 percent from North America.[25]

Canada, India, the Republic of China (Taiwan), and Hong Kong were the top four senders of students to the United States in 1969–1970, with Iran coming in fifth. Thirty-one countries were represented by 1,000 or more students. The 134,959 foreign students in the United States in 1969–1970 were hosted by 1,734 institutions.[26]

The field-of-study category breakdown was almost identical to that of preceding years. Twenty-two percent of all foreign students were studying engineering; in second place was the humanities, with 20 percent; physical and life sciences were third, with 16 percent; the social sciences garnered 13 percent; and business administration was studied by approximately 12 percent.

Table 6.1 1969–1970 Countries Represented by More than 1,000 Foreign Students

Countries	Number in 1969–1970	Number in 1968–1970
Canada	13,318	12,852
India	11,327	9,457
China, Republic of	8,566	7,918
Hong Kong	7,202	5,675
Iran	5,175	4,554
Cuba	4,487	4,878
Thailand	4,372	3,586
United Kingdom	4,216	3,607
Japan	4,156	4,123
Korea	3,991	3,765
China, unspecified	3,463	2,637
Philippines	2,782	2,565
Germany, Federal Republic of	2,634	2,512
Mexico	2,501	2,031
Israel	2,288	2,079
Colombia	2,045	1,833
France	1,977	1,625
Nigeria	1,851	1,790
Greece	1,811	1,635
Venezuela	1,722	1,556
Pakistan	1,576	1,399
Jamaica	1,353	1,392
Brazil	1,343	1,169
Turkey	1,309	1,236
Peru	1,301	1,171
Italy	1,174	1,064
Argentina	1,079	985
Australia	1,077	1,042
Saudi Arabia	1,029	1,057
Lebanon	1,020	921
United Arab Republic	1,015	836

Source: *Open Doors*, 1969–1970

California, New York, and Illinois, in that order, enrolled the most international students in 1969. Michigan was edged out of the third-place position it had held for the previous two years.[27]

More than 46,000 of the foreign students studying in the United States in 1969 were self-supporting. Similar to previous years, students from the Near and Middle East (40 percent) and the Far East

(35 percent) were most likely to be wholly self-supporting. Approximately 25,000 foreign students were funded in whole or in part from U.S. colleges and universities; private organizations provided financial assistance for about 10,000; the U.S. government helped fund about 7,000 students; and foreign governments were providing aid for an estimated 6,500.

In 1969, IIE also produced one of the first counts of foreign students enrolled in community colleges. This census was noncomputerized, and IIE considered it only a "rough count," but the estimated total

Table 6.2 1969–1970 Foreign Student Enrollment: Top Ten Institutions

Institution	Foreign Students	Total Students	Percentage of Total
New York University	4,182	44,401	9.4
Miami–Dade Junior College	3,998	29,375	13.6
University of California, Berkeley	2,904	28,088	10.3
Columbia University	2,796	24,028	11.6
University of Illinois	2,490	51,926	4.8
University of Wisconsin	2,402	65,257	3.7
University of California, Los Angeles	2,197	30,593	7.2
University of Minnesota	1,875	50,415	3.7
University of Pennsylvania	1,699	19,021	8.9
University of Washington	1,663	32,749	5.1

Source: Open Doors, 1969–1970

Table 6.3 1969–1970 Foreign Student Enrollment: Top Ten States

State	Number of Foreign Students	Percentage of Total (134,959)
California	22,170	16.4
New York	17,701	13.1
Illinois	7,795	5.8
Florida	6,939	5.1
Michigan	6,774	5.0
Massachusetts	6,352	4.7
Pennsylvania	5,248	3.9
Texas	4,902	3.6
Ohio	4,121	3.1
District of Columbia	3,949	2.9

Source: Open Doors, 1969–1970

the institute reported was 13,003 foreign students enrolled in 486 two-year institutions. IIE began producing more exact numbers for associate institutions the following year.

Community Colleges and Foreign Students

Given the attractions of lower tuition and less-stringent admission requirements, community and junior colleges began to enroll expanding numbers of foreign students during the 1960s and 1970s. Typically, students intended to conserve money for the first two years and then transfer to a four-year institution.[28] The majority of internationals opting for junior colleges at the time were unmarried males from developing countries, with plans to eventually earn degrees in engineering or business.

This distinctive population, which was beginning to assume importance for junior and community college campuses, had not yet attracted the attention of many researchers; international education and education exchange had not been a focus of community colleges. The earliest writings and studies on junior and community colleges, such as Leonard V. Koos's 1924 report, *The Junior College*, did not address the two-year colleges' potential role in international education or foreign students.[29] Since William Rainey Harper's academic reform initiatives in 1892, the two-year institution had been differently focused than the four-year institution. Harper's bold innovations in Chicago had resulted in the division of the traditional university into two parts—the first to be known as the junior college or academic college, where "the spirit would be collegiate and preparatory, and the second to be known as the senior college or the university college, where the spirit would be advanced and scholarly. . . ."[30]

The localized junior college Harper had imagined in the nineteenth century had taken on an international reach in the twentieth, and American higher education was taking notice. In 1971, the IIE produced its first separate tabulation on foreign students enrolled in community colleges.[31] That year, a total of 15,363 (or about 11 percent of the total foreign student enrollments in U.S. higher education) were enrolled at two-year institutions. According to IIE's 1971 report, 459 two-year institutions reported that they were hosting foreign students. Some striking differences were revealed between the population of foreign students in junior and community colleges and the foreign student population as a whole. Forty-five percent of foreign enrollments in two-year institutions were from Latin America, compared with a national percentage of about 20 percent, for example. Seventeen percent of two-year foreign enrollments were from the Near and middle East, and 22 percent were from the Far East.

Also according to IIE's 1971 report, 24 percent of foreign students in two-year colleges were enrolled in engineering, 23 percent in business administration programs, approximately 20 percent in the humanities, and eight percent in medical sciences.[32]

Research in the field to accommodate the increasing populations during the 1970s was just beginning to appear. Among the contributions were *To Transcend the Boundaries* (1977), a colloquium paper by Edmund J. Gleazer Jr., which discussed why foreign students were choosing community colleges, and *Profile of Foreign Students in United States Community and Junior Colleges* (1977), by Theodore Diener, which studied foreign students' characteristics and demography. *Constraints and Issues in Planning and Implementing Programs for Foreign Students in Community and Junior Colleges*, by S. V. Martorana, and *Effective Programming for Foreign Students in Community and Junior Colleges*, by A. Hugh Adams, also were among the new contributions. All were topics of discussion at the 1977 colloquium sponsored by the American Association of Community and Junior Colleges and the National Liaison Committee on Foreign Student Admissions in Racine, Wisconsin.[33]

The local community had been the focus of junior colleges since their inception, and research related to two-year institutions did not yet include, at least to any significant degree, studies about international enrollees or globally focused ventures. In addition to a review of papers, a search for literature was conducted in preparation for the 1977 colloquium, which uncovered two studies:

The first was a statewide report on international students in public two-year colleges in Florida.[34] Produced in 1972, the study used the Michigan International Student Problem Inventory to determine matters of concern to foreign students. According to the results, financial matters were the first issue reported by the students. Proficiency in English ranked second, and admissions and advising also were listed as important issues. Other areas of concern dealt with student activities, social-personal relationships, living and dining facilities, orientation, health services, and religious programs. More males than females were surveyed for this report, but the concerns varied little according to gender. The report further revealed that the concerns indicated by international enrollees did not appear to diminish for those who had been on campus for some time. Foreign students on campus for 12 months or longer perceived problem areas in about the same way as did the recently enrolled students.

A second study, conducted in Texas public community colleges in 1974, drew similar conclusions and revealed the same student concerns and similar priority of those issues.[35] This report indicated that English proficiency was perceived to be the main problem, followed closely by financial issues.

One of the primary questions posed by those working with community colleges during the late 1960s and early 1970s had to do with student success rates. Just how well did foreign students who attended two-year institutions and then transferred to four-year colleges or universities actually do? Little research had surfaced, in spite of repeated calls for new studies by NAFSA's Community/Junior Colleges Committee.

One of the responses to the call was from Texas Tech University. A study conducted in 1975 revealed some surprising information. The study had concluded that foreign student transfers performed better than carefully screened foreign students direct from high school, domestic students, or foreign student transfers from other four-year institutions.[36] When the study was reviewed a year later, in 1976, however, those who had been involved in the initial report officially added that they believed the situation was in a state of change. New and more flexible institutional acceptance criteria and improved performance by transfers from four-year institutions indicated, they correctly contended, that two-year college students would not be achieving quite as well by the end of the 1970s.

Two-year institutions were nevertheless invested in "going global" and professional organizations were responding. Between 1969 and 1971, a variety of international activities were being conducted by the American Association of Community and Junior Colleges (AACJC), including an international assembly. During the same period, with funding from the Kellogg Foundation, the AACJC established an Office of International Programs.

An important development in 1976 was the formation of the Consortium for International Education (CIE), which established study-abroad programs and provided the basis for the creation of Community Colleges for International Development (CCID).[37] CCID was introduced by Maxwell King and Robert Brueder of Brevard Community College. Before the end of the same decade, CCID had developed the organizational structure and operational philosophy that would support the continued development of international links among community colleges. By 1980, it had established two major overseas clients, the Republic of China (Taiwan) and the Republic of Suriname.

Not all community college educators during the 1970s were in favor of establishing international links, however, or of bringing significant international populations to their institutions. Some believed an international focus was contrary to the original and enduring goals of junior and community colleges. Writing in 1978, S. V. Martorana questioned the validity of increasing international enrollments:

> ... We need to examine the basic assumption that proponents of more action on behalf of foreign students apparently must accept. The assumption has two parts: first, that community and junior colleges can, in fact, attract and enroll larger numbers of foreign students and in larger proportions of their total enrollments, and, second, that a definite course of action to do this should be promoted. How would we answer a friendly critic who might ask: Can they, indeed? And if so, should they?[38]

These questions centered on community colleges' institutional capacity to accommodate more foreign students, as well as their ability to attract, enroll, and generally serve such students' educational and social needs. While most educators at the time agreed that capacity was not a problem, there was considerable debate about the issue of capability. Were the faculty and staff of most community colleges fully prepared to offer international students an outstanding educational and campus experience?

Few denied the importance of the development of international understanding, but according to Martorana, there was the question of whether or not increasing the population of internationals on campus was the right approach for community colleges:

> It must be recognized ... that an international dimension can be developed effectively without bringing more students from other lands to the campus. An international dimension might mean the opportunity for travel abroad by American students, some special co-curricular activities such as lectures, forums, cultural events, and the like, and no increase in foreign student enrollment. Although the effectiveness of such an approach can be debated, it is incontestable that it can be attempted without changing the mix of American and other students on the campus.[39]

Emerging Research

Community colleges were not alone in their need for new and broader research about foreign students' needs and experiences in America. With international enrollments building at every level of higher education, those in the field began to examine foreign students in more detail. In particular, questions about foreign students' adaptation to American culture, and to the academic environment, seemed to predominate.

A literature search in 2001 provided a cross section of studies published during the 1970s (and, occasionally, earlier than that). I. T. Sanders and J. G. Ward's *Bridges to Understanding* (1970), Helm's *Cultures in Conflict: Arab Students in American Universities* (1978), and Pruitt's *The Adaptation of Africans to American Society* (1978) were among the early publications to point out that students have deep-seated attitudes stemming from religion or tradition. Remaining appropriately loyal to those attitudes and beliefs has presented special challenges for international students in their struggle to cope with and effectively function within a new society. Devout Muslims, for example, pray five times each day. Working out times and places to do so is difficult when one also is trying to meet a daily class schedule that was not designed to include Islamic rituals.

W. Frank Hull's study, also during the 1970s, of foreign students centered on their ability to deal with a new environment. Like other researchers, Hull found that foreign students from less-known areas and those most easily distinguishable as foreign by physical characteristics or differing attitudes, were the ones who were less able to integrate during their stay in the States. Hull found that the experience of being foreign in the United States is frequently a difficult and unsettling one.[40]

Because most students' primary reason for being in the United States was to succeed academically, according to Singh (1976), their adjustment to the learning environment was the most critical area of concern. Symptoms resulting from the various sources of culture shock, according to Shogren and Shearer (1973), could manifest themselves in hostility toward the host culture, excessive concern about personal welfare, or fear of being cheated in some way. Problems included social withdrawal, sleeplessness, depression, academic maladjustment, financial problems, loss of self-esteem, or difficulty communicating. Spaulding and Flack (1976) concluded that the problems of foreign students varied depending on the country or region of the world from which they came.[41]

Early 1971, Sharma reported that students from South Asia had better academic adjustment and success than those from either the Far East or Latin America. Chongolnee (1978) agreed that Asians, in general, achieved better overall academic performance than others. Evidence about African students also was mixed: Very early studies by researchers such as Hountras (1956), reported that Africans had the fewest problems. Twenty years later, however, Stafford found that African students had the most difficulty and experienced the highest level of unfriendliness from their respective communities.

Pusch's (1979) introductory essay in *Multicultural Education* was a good overview of the field of intercultural communication. Studies published a few years earlier, such as Conden and Yousef's *An Introduction to Intercultural Communication* (1975) and Cole and Schribner's *Culture and Thought* (1971), offered chapters on value orientations and nonverbal communication as they applied to interactions between different cultures. From the opposite viewpoint, Stewart's *American Cultural Patterns: A Cross-Cultural Perspective* (1972) helped practitioners see Americans from foreign students' perspective. A study by Gullahorn and Gullahorn a few years earlier, in 1963, reported that foreign students returning to their home countries underwent a re-acculturation process similar to that experienced when they began their U.S. sojourn.[42]

The 1979–1980 Foreign Student Census

The Institute of International Education's 1979–1980 census reported 286,340 foreign students from 185 countries and territories enrolled in American higher education—the latest numbers in a twenty-six–year period of continuous enrollment growth.[43] During that period, however, there were numerous fluctuations in the proportionate numbers of students from different regions. Between 1954 and 1979, the percentages of Asians and Africans steadily increased, while the percentage of Europeans decreased.

During the late 1970s, students coming to America from the Organization of the Petroleum Exporting Countries (OPEC) nations were increasing in numbers at a considerably faster rate than students from other countries.[44] This increase was largely attributable to the countries of Iran, Venezuela, and Saudi Arabia, which, combined, supplied 79.1 percent of the increased enrollment from all OPEC countries.

Table 6.4 Number of Foreign Students from OPEC Countries for Selected Years, 1954–1979

Country	1954–1955	1969–1970	1979–1980
Algeria	1	45	1,720
Ecuador	166	648	1,000
Gabon	0	3	0
Indonesia	153	683	2,440
Iran	997	5,175	51,310
Iraq	650	512	1,220
Kuwait	0	319	2,670
Libya	4	286	3,030
Nigeria	268	1,851	16,360
Qatar	0	17	630
Saudi Arabia	40	1,029	9,540
United Arab Emirates	0	2	740
Venezuela	882	1,722	9,860
Total Foreign Student Enrollment	34,232	134,959	286,343
OPEC as a Percentage of Total	9.2	9.1	35.0

Source: Open Doors, 1979

The period from the mid-1970s until the early 1980s marked the peak of Middle Eastern student enrollments in the United States. The spike in oil prices on the world market following the second Arab-Israeli War had caused an abrupt groundswell in student enrollments from OPEC nations. Using profits from petroleum exports, many of these countries were creating generous scholarships that sent students abroad, most of them to the United States.

> They view their American education as an exportable commodity. They come, they buy it, and they take it away. . . . Take the University of Texas, for instance, where many of the 2,000 foreign students are studying petroleum engineering. When it sponsored an alumni conference on energy a couple of years ago, one 1947 grad came a long way back: Sheik Abdullah Tariki, a former Petroleum Minister of Saudi Arabia and a founder of OPEC.
>
> *Time,* 1979

In 1979, Iran was the leading country of origin of international enrollees; approximately 50,000 Iranians were studying in the United

States that year. Nigeria and Venezuela were close behind, and Libya also sent substantial numbers of students. The Gulf States, Kuwait and Saudi Arabia in particular, experienced unprecedented profits from oil and began sending more of their young people abroad for their education. Indeed, much of the surge in the international education market at the time was fueled by petrol dollars.

The Iranian Revolution of 1979 and the seizure of American hostages in 1980, however, changed everything. By 1983 the number of Iranian students in the United States had dropped to around 27,000. Diplomatic relations in the years that followed continued to sour, particularly between the U.S. and several Middle Eastern and North African nations—mainly Iran, Libya, and Syria. The government of Libya, citing economic as well as political reasons, ended fellowships for study abroad and by 1985 all Libyans studying in the U.S. had been returned home.

> "The spirit of seeking understanding through personal contact with people of other nations and other cultures deserves the respect and support of all."
> Gerald R. Ford, 1976

Further complications in the 1980s, such as the suicide bomb attack on the U.S. marine barracks in Lebanon in 1983 and the downing of Pan Am Flight 103 over Lockerby, Scotland, in 1988, and a string of other incidents further strained relations between the United States and the Middle East, serving to discourage Arab students from enrolling in American institutions.

These complications and strained relations did not affect only Middle Eastern enrollments, however. Increasing restrictions and requirements for incoming foreign students at the border, and a "less inviting" attitude toward nonresidents in general, encouraged more international students and scholars to look to other countries to fulfill their higher-learning ambitions.

7

The Late Twentieth Century

Global Competition

Near the end of the twentieth century, there was growing concern on the part of American educators (and politicians) about new and robust competition from other countries for foreign student enrollments. In 1995, The *Chronicle of Higher Education* reported that Australian universities were forecasting a five-fold increase in fee-paying foreign students over the next fifteen years.[1] Based on a study by the International Development Program, the overseas marketing arm of the Australian Vice-Chancellor's committee, the report went on to say that the enrollment gain was predicted to add an estimated four and a half billion U.S. dollars a year to the Australian economy.

Not only Australia, but two other English-speaking nations—Britain and Canada—were directly and aggressively competing with the United States for foreign students. All three of the former significantly increased their recruiting efforts, marketing their institutions abroad through information offices and state-sponsored campaigns. Australia was especially focused on Asia, and presented itself as an affordable alternative to study in the United States. The Australian dollar had declined in value against the U.S. dollar by almost a third by the end of the 1990s, making the Australian's presentation especially alluring to cost-conscious Asians. Australia further enticed prospective students with promises of fewer immigration regulations, which would make foreign students eligible for permanent residency immediately after graduation. All of this coincided with a period when America's borders were tightening.

In September 1998, the Director of the United States Information Agency (USIA) and the President of Educational Testing Service convened a summit at the State Department to assess the situation.[2] Participants in what came to be called the Summit on U.S. Leadership in International Education included representatives from institutions of higher education, U.S. corporations, nonprofit organizations, and government entities. With the help of USIA, the conference participants sought to identify barriers to international educational exchange between the United States and other countries and to formulate a plan for maintaining America's competitive position.

By the end of the conference, the participants had concluded that intensified competition from other English-speaking countries was only one of several reasons for the erosion of America's position in the world of international study. Writing for the conference, Dr. Ted Sanders, President of Southern Illinois University, identified various factors, including complacency on the part of U.S. institutions of higher education in promoting themselves to foreign students; the failure of state and federal governments to facilitate a robust and well-coordinated spirit of entrepreneurship in international education; and diminishing federal funds to support overseas educational advising centers (affiliated with the USIA). Another concern was the lack of coordination at the federal level between the State Department and the Immigration and Naturalization Service (INS), as evidenced by burdensome visa regulations. Many INS officers had overly heavy caseloads, resulting in visa interviews that sometimes shortchanged foreign applicants. "If we are to regain our position of dominance in this very important area," wrote Sanders, "we must now begin to emulate the enlightened policies of other advanced nations who have seen the future and are aggressively pursuing it. Nationally, we must enhance our tangible support for international efforts within a framework of a broad-based, clearly defined strategy. . . ."[3]

It was generally agreed among the conference participants that government officials at all levels had contributed to this disturbing trend. Dr. Sanders continued, writing that in years past, the United States had relied heavily on its overseas educational advisement centers (supported by the USIA) to communicate the strengths of American higher education to native populations. Federal funds to support these centers, however, were steadily diminishing, forcing some of the centers to eliminate staff or shut down completely. Cuts in federal funding also had affected Fulbright and other exchange programs. At the state level, the conference report declared a need for alliances between the various administrative areas—such as state university boards, state governments and commerce officials—in order to reinforce effective planning and promotion of education exchange.

In addition to their conclusions and recommendations, conference participants raised the following points:

- In 1997–1998, the nearly 500,000 foreign students in the United States had contributed more than $8 billion to the U.S. economy.
- Foreign students in Australia had contributed more than $1 billion to the Australian economy, and foreign students in the United Kingdom contributed approximately $2 billion to that country's economy.
- Distance-learning technology was creating new outlets for the marketing of education around the world. In some cases, students could earn a degree from a foreign university without ever leaving home.
- Many educational systems around the world—in India, for example—were strengthening their capacity and increasing enrollments in order to keep their "best and brightest" at home for their higher education.
- The relative strength or weakness of the economies of other countries relative to that of the U.S. affected the ability of students from those countries to afford schooling in America—as evidenced by the mid-1990s financial crisis in Asian countries.
- The ability of foreign students to study in the U.S. was further inhibited by a more complex regulatory environment than most of the competing countries.[4]

Conference attendees were in support of a clear federal policy on international education. They suggested also that an alliance in support of international education be created; that a public awareness campaign should be implemented; and that a comprehensive and extensive marketing strategy be developed for the promotion of American higher education.

The Community College Alternative

"For foreign students, the path to an American college degree increasingly runs through a two-year institution," declared Institute of International Education (IIE) president Allan E. Goodman.[5] When he made that statement in 1997, two-year colleges were enrolling twice as many foreign students as they had only a decade before, accounting for 15 percent of all foreign enrollments in the United States. While international enrollments at all U.S. institutions had grown by only about 7 percent from 1993 to 1997, they had grown a full 20 percent in two-year institutions, according to the 1997 IIE foreign student census. By the end of the century, community colleges had the largest foreign student enrollment percentage increase of any institution type. Associate-degree institutions were now the third-largest type of host, following research-focused institutions and master's institutions. Community colleges in California, Florida, New York, and Texas were enrolling the largest numbers of foreign students.[6]

Continuing a trend that began in the 1970s, foreign students, looking for educational opportunities in the United States that met both their interests and budgets, were turning in ever-increasing numbers to two-year colleges. International students were attracted to these colleges for many of the same reasons as were American students. Associate-degree institutions provided a low-cost, quality education, with flexible and innovative programs of study not found at traditional four-year colleges and universities. In addition, admissions standards and English requirements were typically not as rigorous as in four-year institutions. Some foreign students elected to begin their American education in a smaller, more rural environment (where most of America's community colleges were located at the time), to ease their adjustment to the new systems. Some internationals were specifically seeking intensive English training, and others needed short-term training in a specialized field—but for most, the two-year college was expected to serve as a gateway to a four-year institution and a university degree.

A contributing factor to the enrollment increases at community colleges was the brief period of economic turmoil in the mid-1990s in countries such as Indonesia, Malaysia, South Korea, and Thailand, which had resulted in a general decline in Asian enrollments in U.S. higher education. Some foreign students were now transferring from four-year institutions to associate-degree institutions to save money—so, while there was an overall decline, the community college sector actually increased its foreign student population on many campuses.

Most two-year institutions also attributed increases in international enrollments to the vast amount of information available on the Internet.[7] The majority of America's 1,200 community colleges were maintaining their own Web sites, which were inundated daily with requests for information from potential foreign applicants. Some colleges began including Web pages that were specifically addressed to international students. One fact that potential international applicants were quick to notice was that the annual tuition and fees of public two-year institutions were typically about half that of a public four-year college or university.

For more than thirty years, community colleges had only dabbled in the business of international education, and until the ending decades of the twentieth century, most education exchange efforts had been left largely to the four-year institutions. Only after the 1970s did two-year institutions begin to focus attention on these efforts. Community colleges had been conceived for the purpose of serving their local communities. As the century progressed, however, and technology advanced to the point of enabling the public to communicate and interact globally, the term "community" needed to be redefined. In the case of the two-year college, "community" could no longer refer only to a local geographical area. To serve the local community, these colleges' reach now had to be global.

Factors other than advancing technology had helped to mobilize the community college toward international education.[8] Growing numbers of immigrants, with different traditions, belief systems, and languages, were moving into small towns and rural communities where American citizens had little experience with cultural diversity. Community colleges could provide global education and international experiences for local schools and organizations to help promote a better understanding within their communities of people from different backgrounds. Moreover, the local economies in most communities across America were becoming increasingly dependent on relationships and profitable collaborations with other countries.

In 1988, the Commission on the Future of Community Colleges released a report titled *Building Communities: A Vision for a New Century*. The nineteen-member commission was led by co-chairs Senator Nancy Kassebaum of Kansas and Ernest Boyer, then president of the Carnegie Foundation for the Advancement of Teaching. *Building Communities* was the result of 18 months of study, public hearings, campus visits, and much debate. In it, the commission developed recommendations for the future of community, technical, and junior colleges.

> "Community colleges offer high-quality education at an affordable price, and international students recognize that.... They tend to be quite flexible and offer the types of technical vocational programs and intensive English language training that many seek."
> Todd M. Davis, Institute of International Education

Both the report and Ernest Boyer went on to assume an instrumental role in the evolution of global education for community colleges. Years later, at the 1994 conference, Boyer's contributions provided further impetus for the expansion and institutionalization of global education efforts in the nation's community colleges. To "ensure the survival and well-being of our communities," the report claimed, "it is imperative that community colleges develop a globally and multiculturally competent citizenry."[9]

Another advocate of both community colleges and international exchange was the Stanley Foundation, a private organization working toward a secure peace through the creation of forums, media, and education. In 1994 the foundation, cooperating with the American Council on International Intercultural Education (ACIIE), sponsored a two-day conference, "Building the Global Community: The Next Step," that provided a vision of community colleges' international imperative.[10]

In 1995, in order to gauge the degree to which community colleges in the United States were involved with international education services and programs, the American Association of Community Colleges (AACC) sent a questionnaire to all 1,154 of its affiliates. More than 600 (53 percent) responded. Fifteen percent indicated they did not offer any sort of international experience.

Some of the findings of the survey included:

- Of the 624 colleges that responded, more than 80 percent were offering some type of international education program.
- The reporting colleges had academic or business connection with 143 countries.
- Approximately 80 percent of the colleges had international enrollments.
- States with the greatest numbers of international students in community colleges were California (65,287), Florida (24,146), Washington (14,709), Virginia (11,389), and Texas (11,033).
- Sixty-four percent of the colleges were offering foreign language curricula.

- Approximately 63 percent offered English as a Second Language (ESL) programs.
- Thirty-five percent conducted study abroad or other exchange opportunities.
- Seventeen percent provided international training for local businesses.
- Sixty-one percent of the colleges supported a staff and/or faculty members who were responsible for international education programs.
- Twenty percent of the colleges had received federal funds to help support their programs; 14 percent had been aided by state governments; and private foundations had provided funding for 12 percent.
- Thirty-five percent had collaborations with other international education organizations.
- Twenty-five percent were members of international education organizations.
- Seventy percent agreed that their international programs would expand over the next several years.[11]

The AACC, which conducted the 1995 survey, had been operating since 1920 (initially as the American Association of Junior Colleges), serving as the primary advocacy organization for community colleges at the national level.[12]

In 1996, another conference, "Educating for the Global Community: A Framework for Community Colleges," defined characteristics of globally competent learners. In 1997, a third gathering was held for governmental officials and community college leaders on the topic of "Building Constituencies for U.S. and Community College Involvement in the Global Arena" in an effort, among other things, to clarify public policy issues and identify new vehicles for collaboration between community colleges and governmental agencies. The conference was considered the initial step in a broad-based process of compiling a comprehensive guide to resources for international ventures in community colleges, and for assisting federal agencies with the advancement of their international objectives.

The Stanley Foundation sponsored additional meetings, workshops, and retreats for the leadership of the AACC, the American Council on International Intercultural Education (ACIIE), the Association of Community College Trustees, and Community Colleges on International Development (CCID).[13] It also organized an extensive nationwide program, the Global Community College, to help community colleges begin and develop international curricula, and to enhance international programs on their campuses. The goal for these and other efforts by the Stanley Foundation was to establish viable global education programs in every community college in the United States.

Making the Sale

Recruiting foreign students was not on the agendas of most community colleges until the 1980s. By the end of the following decade, however, two-year institutions across America had moved into the business of attracting and enrolling more internationals. According to Linden Educational Services, an organizer of international student-recruiting tours, representatives of community colleges were among those taking part in all of the tours the company ran in the fall of 1997—throughout Asia, Latin America, and the Middle East.

"We like to recruit so that we really have a mix of international students, but 17 percent are from Europe, and we're trying to increase our numbers from Latin America, Africa, and the Middle East," said Elena Garate-Eskey, dean in international education for Santa Monica College. "International students are very savvy," she added. "If they want an education in the United States, they really shop for it."[14] In the 1990s many students from Japan were enrolling in community colleges. "The Japanese students are much more like the Americans who study abroad—they go for a shorter term, not to get a degree, which is untypical of the foreign-student population at large,"[15] said Peggy Blumenthal, vice-president of educational services at IIE.

The increasingly aggressive recruitment by American's community colleges was being challenged, however, by those who questioned the tactics that some of the colleges were using. Charges of misrepresentation and empty promises by overseas recruiters (who were paid commissions based on how many new students they could enroll) were leveled against a number of institutions. Paying recruiters' commissions, a practice discouraged by the ethics codes of the admissions trade, had led to false expectations on the part of some students, according to some reports. In their eagerness to sell the product, overzealous overseas recruiters sometimes overstated the benefits of the college. Scandals several years earlier, in which trade-school recruiters allegedly signed up the homeless straight out of the shelters, had prompted the U.S. government to ban commission-based recruiting of students who qualified for federal aid.

There were increasing numbers of reports from students who had enrolled based on information from the overseas recruiter, only to find that the college did not turn out to be appropriate for their goals or needs. "Community colleges are often more interested in income than enrolling student who are a good match,"[16] said Linda Heaney, president of Linden Educational Services. A case in point was the story of two students from Hong Kong. After failing their college entrance examinations, each paid $250 to a recruiting agent to be placed in an American college. The community college represented by the agent expedited the students' admission by waiving its requirement of a score of at least 500 on the Test of English as a Foreign Language. Based on the college's own test, the students were placed in an upper-level English as a Second Language class. One failed and one received a "D." At the time of the publication of the article that reported the situation, both women had given up on their plans for earning the two-year degree and then transferring to an elite university.

Some community colleges reportedly were buttressing their recruitment by misrepresenting their record of student transfers to other schools. An article in the *Wall Street Journal* presented two such examples.[17] One described a California community college that had acknowledged its Web site claim that it had an "articulation agreement" with Stanford University was wrong. Typically, articulation agreements mean that a college is adhering to a specific set of academic standards in coordination with the cooperating institution. Upon completion of the two-year program, according to the alleged articulation agreement, students would be guaranteed admittance, or at least given priority in admittance, to the higher-level institution. Stanford said it had no such agreement with that institution.

Study in the USA, a widely read recruitment publication, cited a number of two-year colleges that advertised the same sort of agreement with Harvard, the Massachusetts Institute of Technology, and Boston University, among others. Again, upon investigation, none of those institutions had records of any student transfers from the cited colleges.[18]

Community colleges have wanted to recruit internationals aggressively and bolster their foreign enrollments for several important reasons. Creating more "international" campus environments was promoted as a chief motivation, but economics were a factor, too. First, the majority of foreign students (approximately 80 percent) were self-funded, making them potentially more profitable (and easier to process) than other student populations. Second, they paid higher tuition than in-state students. Third, community colleges in California and most other states were able to keep those extra funds on campus, while tuitions collected from local residents went directly to state coffers, to be redistributed throughout public higher education.

Foreign students who enrolled in community colleges during the 1980s and '90s did not always migrate to obvious destinations. For example, one might assume that two-year institutions in Texas would have attracted a large number of Mexican students to their campuses. While many Mexican and Latin American students did, indeed, enroll at these locations, some institutions, such as Tarrant County Junior College in Fort Worth, enrolled far more students from Africa, India, Sweden, Greece, Canada, and the Philippines.[19]

Despite the problematic methods employed by some two-year institutions, by the end of the century, community colleges as a group were excelling at the task of recruiting and hosting foreign students. For growing numbers of internationals, the two-year experience was the right one. With most prognosticators forecasting even greater international enrollments in two-year institutions, at the end of the century the two-year alternative had become an important option for international students. Community colleges were beginning to play a significant role in America's quest to maintain its position as a leading destination for foreign students.

A Closer Look at International Students

As foreign student populations grew nationwide, the institutions and educators serving these students' needs sought more information about them. What were the distinctive characteristics that practitioners needed to know and understand if they were to do the best job of recruiting, admitting, socializing, involving, and educating these students?

Institutions can play a major role in facilitating the students' adjustment to life in a new culture. Foreign students' most commonly reported concerns include homesickness, finances, housing and food, English language proficiency, adjusting to the American classroom environment, preparing written and oral reports, understanding American customs, making friends, forming relationships with members of the opposite sex, and being accepted by social groups.[20] The major variables affecting student adjustment have been shown through a variety of studies to include national origin, undergraduate versus graduate status, gender, marital status, and major field of study. Many institutions have worked to address these problems by providing on-campus intensive English classes, international student organizations, special activities, and host family programs.

This process could not be accomplished efficiently, however, without information; not surprisingly, the new demands of hosting more foreign students encouraged an increase in studies that pertained to them. Many of these studies focused on adjustment and academic performance, some on social and psychological impact, and others on cultural comparisons. More rarely, studies looked specifically at the attitudes and characteristics of foreign students with the intention of discovering how internationals, as a group, differed from the general student population.

Several efforts to compile comprehensive sources for research related to international students took place during the 1980s, such as the one produced by Philip Altbach, David Kelly, and Y. Lulat in 1985.[21] A follow-up by Philip Altbach and Jing Wang was published four years later.[22] Another review of research on foreign students was produced by Paul Marion in 1986 and published in *New Directions for Student Services*.[23]

Most early studies agreed that the academic achievement of foreign students was affected by their attitudes and adjustment. Studies focusing on adjustment also generally agreed that an international student's immediate needs usually related to the process of getting settled, becoming accustomed to the campus climate and the surrounding community, joining appropriate campus activities, and adjusting to new academic practices and expectations. These reviews also noted that it took different amounts of time and varying approaches for internationals to make positive adjustments, and that adjustments were influenced by any combination of an array of factors, such as English language ability, personality characteristics, previous foreign travel, financial status, academic success, cultural differences, race, foreign appearance, living arrangements, age, length of stay, graduate or undergraduate status, preconceptions and expectations, socioeconomic background, and future job prospects upon their return home. Legal and immigration issues could also affect international students.[24]

Complicating the challenge to adjust to American culture were some students' discomfort with being regarded as "foreign" by their classmates. Quite a few studies showed that the students who were most academically well-adjusted were those who had the most frequent interaction with American students and with faculty. Most studies also revealed that foreign students desired close relationships with their professors. Studies also indicated that different learning styles often created obstacles in the academic adjustment process in the classroom.[25]

Gary Althen of NAFSA, writing in *The Handbook of Foreign Student Advising* in 1984, pointed out that whatever their educational differences, most foreign students in the United States found certain aspects of American education to be different from what they had experienced in their home countries. Those aspects that most likely required some process of adjustment for the students included:

- having to select from among a number of possible courses, rather than following a prescribed curriculum.
- being assigned an academic adviser, rather than simply reading about courses that must be taken.
- specializing later, rather than earlier, in the undergraduate program and thus having to take courses outside one's area of interest in order to obtain a "liberal education."
- having to take objective-type tests (such as true-false or multiple choice), rather than, or in addition to, subjective-type examinations (essays).
- dealing with a sometimes complex system for registering for classes each term.
- having relatively frequent assignments and examinations, rather than being left to work independently.
- encountering classmates, especially at the freshmen and sophomore levels, who seemed ill-prepared for, and unmotivated to perform, postsecondary study.
- being expected to raise questions and participate in class discussions, rather than sitting quietly and accepting the teacher's word on all matters.
- encountering competitiveness among students, especially in professional- and graduate-level classes.
- having to analyze and synthesize materials.
- being expected to use the library extensively.
- having a great deal of importance attached to grades.

- having to do what he or she might consider menial tasks in laboratory courses.
- being liable to punishment for activities deemed to constitute cheating or plagiarism.[26]

Studies generally agreed that a foreign student's country of origin was an important factor in the student's ability to adjust. Canadian and Western European students usually had the fewest overall adjustment problems, while Asians experienced the most. This was attributed chiefly to differences in English language abilities, but also to cultural differences between Asians and Americans. Latin and European students tended to be more generally satisfied with their U.S. educational experience, interacted more frequently with American and other students than did those from other world regions, and had the fewest reported problems. Asians interacted the least, and many Asian students (especially those from the Peoples' Republic of China) had a tendency to remain within their own cultural enclaves while on campus.[27]

Regardless of the foreign student's origin, however, the research generally agreed that international students who were well-involved in extracurricular activities, involvement programs, or other forms of interactions with American students, faculty, or students from other cultures, were the most satisfied with their college experience. In Abe, Talbot, and Geelhoed's 1998 study, for example, newly admitted international students participated in an International Peer Program. Results suggested that the participants experienced significantly higher social adjustment scores than nonparticipants.

Attitudes and Characteristics of Foreign Students

One of the few comprehensive studies on the national level that compared international and U.S. student attitudes and characteristics was conducted by the American Council on Education.[28] Based on the results of its 1994 Cooperative Institutional Research Program (CIRP) Freshman Survey, published in the 1996 issue of *Open Doors*, the report enumerated important differences between U.S. and foreign students.

The report revealed, among other things, that foreign students appeared to be better prepared for the collegiate experience than American students in general, with more than 40 percent of foreign freshmen reportedly earning top grades in high school. Additionally, foreign students reported excellent preparation in the sciences, compared with American students. In fact, the only areas where the proportion of American students reported stronger preparation than foreign students were in English and American history.

According to the report, foreign students, more than American students, saw college as an opportunity to become more broadly educated individuals. More than 60 percent of foreign students looked to college as a place to become a more cultured person, as compared with only 36 percent of U.S. students. Americans, on the other hand, wanted a college degree for more practical reasons. More than 76 percent of U.S. students looked to college as a means to a better job and more money.

Internationals generally came from households that were either quite poor or quite wealthy, while most American students were from middle-class backgrounds. Foreign freshmen were less concerned about financing their collegiate experience than American students. Approximately 40 percent reported having no concerns at all, compared to 30 percent of U.S. students.

Reasons for selecting institutions also differed between the two freshmen groups. Americans made their selection not just on academic reputation, but also on the basis of tuition (low) and proximity to home. Foreign students made their college selection largely on the basis of academic reputation (62 percent) and whether or not a school's graduates were accepted into top graduate schools (35 percent).

The occupational aspirations of foreign freshmen were generally higher than those of incoming American students. Foreign freshmen's self-confidence in academic areas was much stronger than that of American students, and a greater proportion of foreign students planned on earning a graduate degree than did Americans.

American students placed greater importance on religion in their daily lives than foreign students. At the same time, U.S. students were more likely to drink alcohol, "hook up" with short-time acquaintances, feel overwhelmed, stay up all night, and fail to complete assignments. Americans saw themselves as being above average, when self-rating, for such characteristics as "stubbornness," "cooperativeness," and "physical appearance."

Reentry

One conclusion of Marion's literature review was that little research had been conducted to determine the needs of foreign students returning home (reentry issues), or about whether American education had met their needs. A study by Hansel in 1993 tracked the readjustment experiences of forty-nine Indian students who returned home after graduating with a U.S. degree. Interviews revealed that most of the students experienced stress or other difficulties, with problems ranging from moderate initial anxiety about getting a job and shock at the crowded conditions, pollution, and poor services to an intense period of depression and alienation. Other literature designed to inform both practitioners and students about reentry issues was country-specific. *Returning to Russia* (1998), published by the American Chamber of Commerce in Russia, was an example of literature that provided statistical data about returnees and advice about the Russian job market.[29]

Upon returning to their home countries, some graduates had an easier time than others in finding an appropriate job—depending on their field. In many parts of Asia, for example, businesses are "computer compliant," and returning students with degrees in computer science from the U.S. could usually find a job right away. Asian students who earned degrees in the arts, education, or other fields, however, had more trouble finding a professional niche where they could appropriately translate their U.S. study to fit their home country's expectations. The studies also showed that former international students who had previous educational experience in the U.S. indicated that a longer stay in America, and more practical experiences in their academic programs, would have made their overall experience more valuable and the process of reentry more manageable. Optional Practical Training, or Curricular Practical Training, often provided welcome opportunities for many internationals to experience the American workforce before either returning home or becoming a permanent U.S. resident.

Intellectual Migration

By the final decades of the twentieth century, foreign-born scientists and engineers were contributing significantly to the brain power of the United States. Looking at the United States' labor force with doctoral degrees in science and engineering fields, immigrants comprised nearly 30 percent of those conducting research and development, according to the National Science Foundation:

> Many decades earlier, the emigration of such highly skilled personnel to the United States was considered one-way mobility, a permanent brain

drain depriving the countries of origin of the "best and the brightest." More recently, however, the mobility of highly talented workers is referred to as "brain circulation," since a cycle of study and work abroad may be followed by a return to the home country to take advantage of high-level opportunities.[30]

The large foreign component of U.S. human intellectual capital was linked to the success of American higher education in attracting, supporting, and retaining foreign science and engineering graduate students. Foreign students, particularly those from Asia, represented a large fraction of enrollment and degrees in science and engineering fields in American graduate institutions. In 1995, of the 420,000 graduate students in science and engineering programs in the United States, roughly 100,000 were foreign students—most from about a dozen countries of origin. In 1995, at the doctoral level, foreign students (including those with permanent and temporary visas) earned 39 percent of the natural science degrees, 50 percent of the mathematics and computer sciences degrees, and 58 percent of the engineering degrees. Students from China, India, South Korea, and Taiwan accounted for more than half of these science and engineering doctorates.[31]

In a letter to the editor of the *Chronicle of Higher Education* in 1999, Jose L. Torres, Dean of the School of Computer Science and Mathematics at Marist College (Poughkeepsie), wrote:

> In truth, American students are not flocking to these [science and engineering] programs, because they tend to believe that the potential rewards of a PhD in science do not justify the effort. The cornerstone of this belief is a lack of interest in math and science, quite possibly born and nurtured way back in elementary school. Who is to blame for this state of affairs? Theories abound, but I find it hard to believe that international graduate students have anything to do with it.[32]

Financial support available from academic research activities was a major factor in attracting foreign students to U.S. doctoral programs. More than 75 percent of the 10,000 foreign doctoral recipients at American universities in 1996 reported their universities as the primary source of support for their graduate training—most in the form of research assistantships. Financial resources for research assistantships were provided to universities by federal government agencies, industry, and other nonfederal sources in the form of research grants. During the same period when academic research expenditures were growing, the number of foreign doctoral students supported by university science and engineering departments was also increasing. From 1985 to 1996, academic research expenditures increased from $13 to $21 billion. The number of foreign doctoral students primarily supported as research assistants more than tripled—from 2,000 in 1985 to 7,600 in 1996.[33]

Between 1988 and 1996, foreign students from major Asian and European countries, Canada, and Mexico earned more than 55,000 science and engineering doctoral degrees. During this period, about 63 percent of these doctoral recipients hoped to remain in the United States after completion of their studies, and about 39 percent had firm plans to do so. The proportion of foreign students who remained in the United States, referred to as the "stay rate," varied widely by country. About half of the foreign doctoral recipients from China and India were accepting firm opportunities for further study and employment in the United States. In contrast, only 23 percent of the doctoral recipients from South Korea and 28 percent from Taiwan accepted offers to remain in America. Foreign science and engineering doctoral recipients remaining in the United States usually did so through postdoctoral study.[34]

Employment offers to foreign doctoral recipients were strongly geared toward research and development, mainly within business or industry. A recent study of foreign doctoral recipients working and earning wages in the United States (Finn, 1997) shows that about 47 percent of the foreign students on temporary student visas who earned doctorates in 1990 and 1991 were working in the United States in 1995. The majority of the 1990–1991 foreign doctoral recipients from India (79 percent) and China (88 percent) were still working in the United States in 1995. In contrast, only 11 percent of South Koreans who completed science and engineering doctorates from U.S. universities in 1990–1991 were working in the United States in 1995.[35]

Enrollments in the Late Twentieth Century

Toward the end of the century, international students accounted for about 3 percent of the 14 million graduate and undergraduate students in American higher education. In spite of the fact that the United States hosted the greatest portion of the world's international students, the numbers seemed small to many in the field. Such was the opinion of

Allan Goodman, academic dean of the School of Foreign Service at Georgetown University:

> Even with the high cost of postsecondary education in the United States, that (three percent) figure is surprising, given the popularity of American institutions abroad and the limited number of seats for college students elsewhere, especially in Asia and Europe. California colleges, for example, have more spaces for students than does all of China; the United States has four million more spaces in its colleges than does the entire European Union. Yet only about ten percent of American colleges have foreign-student enrollments larger than five percent, and only a handful proclaim in their mission statements that they are committed to "internationalism"—that is, to promoting international-affairs courses as part of a liberal-arts core, to increasing opportunities for foreign study by students and faculty members, and to recruiting students from abroad.[36]

Concerns over what some saw as lackluster numbers had been at least partly brought about by a slowing of foreign student enrollment growth since the 1980s. The 1990s saw a reduction of students coming from Asia in particular as a result of economic problems in that region.

The 1999–2000 Foreign Student Census

According to IIE's *Open Doors 2000* report, there were 514,723 international students studying in United States colleges and universities during the 1999–2000 academic year—an increase of 4.8 percent over the previous year.[37] International students represented 3.8 percent of all U.S. higher education enrollments, but were enrolled in much greater proportions at higher academic levels. Foreign students made up 12 percent of all U.S. graduate enrollments. All data and tables in this section are from IIE's 2000 report.

By academic level, 59,830 foreign students were studying in two-year institutions (1.2 percent of the total enrollment of two-year institutions). A total of 177,381 internationals were enrolled in bachelor's institutions (2.7 percent of the total student enrollment); and 218,219 foreign students were studying in graduate programs, comprising 12.9 percent of graduate-level enrollments.

New York University remained the leading host of international students in the United States, followed by the University of Southern California, Columbia, and the University of Wisconsin.

Asian students constituted more than half of international enrollments in 1999 (54 percent) and Europeans were the next largest region represented with 15 percent of international enrollments. Canada ranked sixth among the leading senders, with more than 23,000 students in the United States. Foreign student enrollments at the end of the century reflected substantial increases from China, and especially India, whose enrollments grew at more than twice the overall rate.[38]

Table 7.1 1999–2000 Leading Institutions by International Student Enrollment

Rank	Institution	International Students	Total Enrollment
1	New York University	4,890	37,077
2	University of Southern California	4,564	28,906
3	Columbia University	4,532	21,453
4	University of Wisconsin–Madison	4,154	41,089
5	Purdue University Main Campus	4,133	36,878
6	Boston University	4,126	28,493
7	University of Michigan–Ann Arbor	4,101	37,828
8	University of Texas at Austin	3,992	48,906
9	Ohio State University (Main Campus)	3,880	48,003
10	University of Illinois–Urbana	3,454	36,690

Source: Open Doors, 1999–2000

Table 7.2 1999–2000 Foreign Student Enrollments: Top Ten Places of Origin

Rank	Place of Origin	Number Enrolled	% of US Foreign Student Total
	WORLD TOTAL	514,723	
1	China	54,466	10.6
2	Japan	46,872	9.1
3	India	37,482	8.2
4	Korea, Republic of	41,191	8.0
5	Taiwan	29,234	5.7
6	Canada	23,544	4.6
7	Indonesia	11,300	2.2
8	Thailand	10,993	2.1
9	Mexico	10,607	2.1
10	Turkey	10,100	2.0

Source: Open Doors, 1999–2000

Growing Enrollments from India

"Every year the US [sic] consulates in India give visas to nearly 1,500 students to pursue higher studies in American universities,"[39] said one official at the United States Embassy in New Delhi. According to a 1998 UNESCO report, although India was home to more than 260 universities comprising some 8,000 colleges, many prospective Indian students felt that the standard of higher education had not kept pace with change. "There is no point in doing research in India. The lack of up-to-date facilities and funding forces students to come to the United States,"[40] remarked New York University researcher Virul Acharya.

Only minor changes had been implemented since the educational system was set up during the British era, and educational institutions, most state-aided, suffered from years of federal budget cutbacks. According to Acharya, higher research had been stagnating since the 1970s. "Most Indian colleges do not even have the Internet,"[41] he observed.

With guidance from their predecessors, Indian students carefully selected which institution to apply to, for a degree from a reputable institution would assure them of a lucrative job upon completion. Many were recruited through campus interviews conducted by multinational companies.

> It is a familiar sight to the residents of Madras and other big Indian cities. Oblivious of the scorching sun, scores of young people stand in long queues outside the United States consulate waiting for their turn to be interviewed. Many of them are students wishing to go to their dreamland for higher studies, if possible for a bright future too. While some of them manage to get visas, others fail to convince the consulate officials.
>
> UNESCO *Courier*, 1988

In 1998, American colleges and universities saw an explosion in Indian enrollments. Up almost 30 percent from the preceding year, the total number of students from India in U.S. higher education approached 55,000—now taking second place (after China) among leading sending countries.[42]

For those who decided to return home to India, however, a well-paying job was often hard to find, and few companies in India offered lucrative salaries. "I came here to do my studies in computers and

I got a job the day after the course was over," said Raj Lokaiyan, a computer-professional-turned-businessman living in New York. "I am sure in India I would have had to wait for months, maybe years, to find a suitable job."[43] Two decades earlier large numbers of Indian students were coming to the United States to study medicine. In the 1990s the most popular field was information technology, followed by business administration. According to *India Abroad*, an ethnic Indian weekly published in the United States, there were nearly 35,000 Indian computer professionals in California's Silicon Valley alone.

Most Indian students coming to the United States or England in the 1980s and 1990s were from middle-class families, many of whom were seeking elevated social status by sending their children abroad. In the past, British universities attracted the majority of Indian students because of the historical and colonial ties between the two countries. Drastic cuts in the number of scholarships in Britain during the 1970s, however, had helped to reroute these students' migration toward America—the obvious alternative because of the language similarities and the higher level of education.

Borders and Taxes

The first bombing of the World Trade Center, in 1993, caused a public outcry for the enhancement of governmental systems that tracked international students in the United States. The suicide bomber, Eyad Ismoil of Jordan, had entered the U.S. on a student visa. Ismoil was born in Kuwait, attended high school in Jordan, and came to the United States in 1989 to study engineering at Wichita State University in Kansas. At his trial, Ismoil read from a prepared statement: "Jail me and you will add one number to the wrong list. But don't think that you will ever rest because tyrants always end up in trouble," he said. "In the world, a fair trial is always rare."[44] The judge sentenced the 26-year-old to 240 years.

In part motivated by the World Trade Center attack, President Clinton signed the Illegal Immigrant Reform and Immigrant Responsibility Act of 1996 (IIRIRA) into law.[45] This legislation affected every international student in America, as well as most of the college officials who served their needs. New documentation, tracking, and reporting demands were now mandated.

> "No one who has lived through the second half of the 20th century could possibly be blind to the enormous impact of exchange programs on the future of countries..."
>
> Bill Clinton, 1993

The new law marked the end of a legislative process had begun shortly after the Republican Party assumed majority party status in the House and Senate after the 1994 midterm elections. The final piece of legislation was vast, more than 200 pages, and was divided into six broad sections:

- Title I — Improvements to Border Control, Facilitation of Legal Entry and Interior Enforcement
- Title II — Enhanced Enforcement and Penalties Against Alien Smuggling; Document Fraud
- Title III — Inspection, Apprehension, Detention, Adjudication, and Removal of Inadmissible and Deportable Aliens
- Title IV — Enforcement of Restrictions Against Employment
- Title V — Restrictions on Benefits for Aliens
- Title VI — Miscellaneous Provisions[46]

Among the areas addressed in the new law was the broad subject of gaining control of the country's border, and the immediate impact of the law was a dramatic increase in the number of Border Patrol agents. Twelve million dollars had been allocated for a fourteen-mile triple fence along the U.S. border from San Diego eastward.

Entry and exit from the United States would become generally more controlled as a result of this new legislation. New border crossing identification cards had to be issued and put in use before the year 2000, complete with "biometric identifiers," such as fingerprints or handprints. According to IIRIRA, the Attorney General was now charged with developing an automated entry-exit control system. New pre-inspection stations would be set up in at least five of the foreign airports sending the greatest number of inadmissible aliens to the United States. In addition, the Immigration and Naturalization Service's internal enforcement efforts would be stepped up.

Most critical to foreign students were the new provisions affecting F-1 and J-1 exchange visitors, which created grounds for exclusion for certain "student visa abusers."[47] Under Section 346 and 625 of IIRIRA 96 (1996), students were inadmissible if they sought to attend a public elementary or adult education program (not a college or university program). They were also inadmissible if they sought to attend a secondary school for more than one year without reimbursing the local school district for the full, unsubsidized per capita cost of the alien's school attendance. Those who entered the United States on an F-1 visa to attend a private elementary, secondary, or language school and who then left that school to attend a public education program, were removable unless the twelve-month limit and the tuition reimbursement requirements were met.

Visas

Types of Student Visas:

- F-1 (Student visa). The F-1 visa is for full-time students enrolled in an academic or language program. F-1 students may stay in the US for the full length of their academic program plus 60 days. F-1 students must maintain a full-time course load and complete their studies by the expiration date listed on the I-20 form.
- F-2 (Spouse or child of an F-1 visitor)
- J-1 (Exchange visitor). The J-1 visa is issued for students needing practical training that is not available in their home country to complete their academic program. The training must be directly related to the academic program. The J-1 visa obligates the student to return to their home country for a minimum of two years after the end of their studies in the US before being eligible to apply for an immigrant (permanent residence) visa.
- J-2 (Spouse or child of a J-1 visitor)
- M-1 (Vocational or other non-academic student). The M-1 visa is issued for students attending non-academic trade and vocational schools.
- M-2 (Spouse or child of an M-1 visitor)

Source: U.S. Citizenship and Immigration Services, Department of Homeland Security

---

To adhere to the new law, F-1 and M-1 schools and universities and J-1 exchange visitor program sponsors now had to collect and maintain a substantial amount of information relating to F-1 student visa holders and J-1 exchange visitor program participants, including contact information and visa status. According to IIRIRA mandates, on April 1, 1997, the INS could require F, M, and J visa sponsors to collect fees (up to $100) from the affected aliens. Institutions failing to provide the information would be barred from sponsoring F, M, or J visas.

A new and extensive monitoring system had been called for in the IIRIRA bill, and in order to comply, pilot programs were conducted to establish an efficient method. One of the systems, the Coordinated Interagency Partnership Regulating International Students, was tested in Alabama, Georgia, North Carolina, and South Carolina. The pilot program called for students to obtain identity cards that included a photograph and a print of the student's right index finger.

Most agreed that pilot programs for new and effect documentation systems had to be tested. The existing system was "obsolete, flawed, and ineffective," said Jackie Bednarz of the INS. "For reasons of national security, the INS was asked to assess whether we had a foolproof system for controlling foreign students," she continued. "The conclusion was we did not."[48]

Some campus advisers worried that the new tracking system would have the effect of turning them into INS agents, saying it would be difficult to both counsel international students and scholars, and at the same time regularly report information on them to the immigration service. "It's hard to help students if you are wearing the hat of a policeman," said Jan Sandor, assistant director of international education at the University of Georgia.[49]

Policing students for tax documentation was another governmental record-keeping system that drew controversy from those on campuses who served international students. The United States Internal Revenue Service (IRS) was often accused of imposing unfair burdens on foreign students. A case in point was recalled in a 1997 issue of *The Chronicle of Higher Education*.[50] The story was about Dian, a student from Indonesia, who was pursuing an undergraduate degree from Lewis and Clark College. Dian had no paying job or U.S. income of any kind, and received no financial aid. Her family provided her entire support—tuition, books, supplies, room, board, and spending money—by sending her the necessary funds from Jakarta. Dian logically assumed that because she had no income, she need not file an income-tax return in the United States.

The IRS, in fact, actually required two highly detailed tax forms from every student, even if they had no income whatsoever to report, and failure to return the forms could subject students to fines and get them into trouble with immigration authorities.

The forms themselves were not only confusing, but also sometimes required information that was difficult for the student to obtain, particularly in an unfamiliar environment. The tax forms called for the student's passport number, visa number, taxpayer-identification number, type of U.S. visa, foreign address, and their reason for coming to the United States. In order to secure the tax-identification number, students with no income had to appear in person before an IRS employee, prove their identity, and make an application. Often students waited a full month before receiving the ID number. When they did obtain it, and could finally file a return, in many cases the form contained zeroes on every operative line.[51]

The students were also asked to disclose information about any scholarships or fellowships; the name, address, and phone number of the director of their academic program; and the exact number of days spent in the U.S. during each of the past three years. Included also were questions such as "Enter the number of days in 1996 you claim you can exclude for purposes of the substantial presence test." Needless to say, many international students either filed incorrectly or gave up altogether. Many educators, such as John Bogdanski, professor of law at Lewis and Clark Law School (and former member of the Commissioner's Advisory Group of the Internal Revenue Service)[52] were calling for reformed and simplified systems, denouncing the existing directives as "gobbledygook."

8

The Opening Years of the Twenty-First Century

International Students and the Rising Threat of Terrorism

As noted in the previous chapter, when it was determined that at least one of the terrorists who participated in the 1993 World Trade Center bombing had entered the country on a student visa, a cacophony of demands erupted, calling for closer surveillance of all international students residing within the country's borders.[1] Galvanized by the attack, Congress responded in September 1996 with the passage of the Illegal Immigration Reform and Immigrant Responsibility Act, legislation that mandated the creation of an electronic reporting and tracking system for international students. By June 1997, the Coordinated Interagency Partnership Regulating International Students, or CIPRIS, a federal task force assembled in 1995, began testing a pilot version of its computerized tracking system in cooperation with twenty-one universities in the Southeast.

In June 2000, the congressionally appointed, ten-member National Commission on Terrorism (NCT; established in 1998 after the bombings of two U.S. embassies in Africa and charged with proposing new measures for minimizing the danger of terrorist activities) endorsed the idea of monitoring foreign students in the United States more closely than ever before. Although the NCT devoted only two pages of its sixty-four–page report to the issue of international students, it did include several specific proposals. Among them was using the existing program, CIPRIS, as the model for a new $45 million nationwide tracking system, to be overseen by the Immigration and Naturalization Service (INS).[2] The successor program, called the Student and

Exchange Visitor Information System, or SEVIS, would be developed with modifications from the original CIPRIS database.

The SEVIS program was considerably more ambitious in scope and detail than its earlier incarnation. Mandatory computerized records would now identify the place of residence of all foreign students, the schools they attended, their academic majors, course loads, and expected graduation dates—detailed data to be supplied by every U.S. academic institution admitting foreigners. Furthermore, students themselves would be required to bear part of the cost of the system's operation and updating through a separate processing fee or surcharge levied when seeking admission to study at a U.S. institution of higher learning.

"Of the large number of foreign students who come to this country to study," the NCT report observed, "there is a risk that a small minority may exploit their student status to support terrorist activity." The narrative continued with considerable understatement, "The United States lacks the nationwide ability to monitor the immigration status of these students." It was further noted that many students admitted for study in the United States were coming from countries such as Iran and Syria, nations officially designated as state sponsors of terrorism. Much was made of the revelation that it was possible for a terrorist to drop out of school yet remain in the country illegally. "Today," the report stated, "there is still no mechanism for ensuring the same thing won't happen again."[3]

Opinion quickly divided. L. Paul Bremer III, the commission's chair, defended the group's intent by claiming that partial measures were preferable to none at all. He conceded, "We're under no illusions that anything you do with foreign students makes a serious dent in the security of a country's borders." But he likened the proposed tracking system to a home security system in which a home owner installs locks on the doors and bars on the windows. Each, according to Bremer, provides an additional degree of security. Similarly, closer surveillance of foreign students, he alleged, might reduce the potential threat posed by hostile interlopers disguised as students.[4]

Opponents of the plan stressed its probable high cost, time consumption, and the intrusiveness into and potential violation of students' privacy entailed. Marlene M. Johnson, executive director of NAFSA: Association of International Educators, vented her irritation that the commission had not provided new data on the potential threat represented by foreign students, apart from the World Trade Center example that seemed to be "trotted out" annually in later years to enflame sentiment in favor of more stringent defensive measures. "We all need to be concerned about terrorism," she allowed. "But we ought not to think we can solve the problem by overtracking and overregulation of people who are not the problem."[5]

Many disagreed, claiming that surveillance was a necessary evil in today's world, and that filling out forms and even paying a tuition surcharge, although inconvenient, was warranted if it afforded all parties a greater measure of security against terrorism. Some critics recalled that in the aftermath of the 1995 Oklahoma City bombing, investigators had been unable to obtain the names of students from states that supported terrorism who were known to be living in the immediate area and who might have turned out to be suspects. The lesson to be learned, defenders alleged, was that the country needed to have in place a more accurate tracking system of all foreign students, no matter what their nation of origin. Others disagreed just as vehemently, arguing that students should not be singled out for closer scrutiny than any other segment of the population, especially as their entry visas collectively represented no more than a tiny percentage of the total number of entry visas granted in any given year.

9/11: The Debate Continues

Whereas the first attack on the World Trade Center resulted in half a dozen fatalities, the larger-scale destruction of the Twin Towers on September 11, 2001, killed thousands. In the weeks that followed, calls went out demanding once again that the Immigration and Naturalization Service accelerate the installation of its highly touted but long-delayed computerized tracking system.[6] Senator Dianne Feinstein of California, for one, urged a six-month moratorium on new student visas and the allocation of $32 million to allow the INS to get the student identification system up and running.

Meanwhile, in the weeks and months following 9/11, bitter recriminations were heard across the entire political spectrum.[7] Stringent entry requirements and new, time-consuming bureaucratic barriers—almost all of them erected in the name of national security—were proposed repeatedly, as was legislation mandating more restrictive visa issuance policies. Which measures promised to enhance public safety and best safeguard the nation's security remained an open question. But in the midst of all the turmoil and confusion, a few points were unmistakable: that the welcome mat for foreign students was being withdrawn; that Fortress America had become a far less hospitable academic venue than ever before for foreigners; and that prospective students should now consider looking elsewhere for a place to pursue their studies.

Demands for concrete action of some sort took on renewed credibility when investigations revealed that again, as in 1993, one or more of the nineteen hijackers in the aerial assault on the World Trade Center complex in 2001 had been in the United States on a student visa.[8] Worse yet, to the mortification of U.S. immigration officials, it was reported that fully six months later, student visa approvals were actually sent to the same Florida flight school where two of the hijackers has been enrolled as they were preparing themselves for the attacks. (Both had held tourist visas during the period they acquired their flight training.)[9]

Notwithstanding this information, many educators nationwide harbored serious misgivings about calls for greater scrutiny of foreign students on American soil. The more extreme proposals, such educators alleged, would accomplish little except to staunch the flow of foreign students to the United States—and the billions of dollars their presence introduced into the American economy.

Many questions remained unanswered. Would law enforcement officials now begin profiling Middle Eastern students exclusively? Was it appropriate or necessary to demand that students themselves underwrite the cost of a national tracking system? Would larger fees and a longer visa-application process depress international exchange and serve basically to discourage students from coming to the United States to study in the first place? Above all, given America's desire to preserve an open society, would restrictions accomplish the goal of protecting the homeland without violating individual civil rights?

Opponents of tracking made it clear where they stood. Enacting draconian measures at the expense of keeping open borders, insisted Li-Chen Chin, director of international programs at Bryn Mawr College would be "extremely shortsighted." Education, he declared, was the best way to promote international peace. Restricting the usual influx of foreign students, Chin concluded, "would not fix the terrorist problem."[10] David Ward, president of the American Council on Education, adduced other grounds for resisting the development of a massive federal database on students, one inevitably susceptible to abuse. Although he was not entirely opposed to better tracking, he felt that federal officials should not depend on schools to gather security-related data if in fact that task proved necessary. "That shouldn't be the university's responsibility," Ward declared. "The university is qualified to make academic judgments about students, not security judgments."[11]

Most objections to new regulations stressed pragmatic considerations and the deleterious economic consequences of any major decrease in international student enrollments. A decline would be painful, it was claimed, because most foreign students typically paid full tuition. Hence, the loss of tuition revenue could be massive if fewer overseas students chose to come to the United States to finish their education.[12] Yet another concern was that fewer students might threaten the existence of certain short-term academic programs such as summer- or single-term intensive English.

But programming worked both ways. Without foreign students, some courses taken by small numbers of U.S. collegians might also be endangered. "International students," alleged Martyn Miller, Temple University's director of international student services, "are providing the fiscal base to offer programs to our American students."[13] Without foreign students, business engineering, computer technology, mathematics, and other programs, courses not attracting large enrollments from U.S. students, Miller claimed, might have to be cut drastically or even eliminated at some schools.

Economic factors aside, however, the most frequently advanced declaration on behalf of the importance of hosting foreign students emphasized the building of bridges across cultures and societies. "A lot of the best friends of the United States," insisted Jim Lynch, director of the international student program at Penn State, "around the world are people who were educated here. They're grateful for their education and they learned what America is all about."[14] Anything that hampered or restricted opportunities for foreigners to become better acquainted with American culture, he emphasized, should be avoided to the greatest extent possible.

Debate continued unabated in the months following the destruction of the World Trade Center and the attack on the Pentagon. The basic arguments advanced on all sides differed little from those put forth after the 1993 terrorist assault. University officials and lobbyists—at least a vocal segment thereof—continued to express their apprehension over the prospect of new regulations that might unduly limit the entry of foreign students into the United States. College lobbyists were especially upset over the government's plan to collect a $95 surcharge from foreign students. Whereas the revenues generated would help finance the national information-gathering system, it was conceded, the practice would send a bad message by singling out international students and discouraging many of them from applying to come to the United States. "The natural reaction would be to close our doors to foreign students, and that would be exactly the wrong step," declared Terry W. Hartle, senior vice president for government and public affairs for the American Council on Education, an advocacy group acting on behalf of most of the nation's colleges and universities. "The lesson [of 9/11] is we need more engagement with the world, not less."[15]

NAFSA: Association of International Educators had long been the leading critic of the proposed student monitoring system. Restricting student visas and tracking foreign students from the time of their arrival on American soil until they left, NAFSA representatives argued, would accomplish little—though they admitted it would be difficult to convince Congress that SEVIS or a similar system was not needed, given the public's demands for decisive action. All the same, however, security would not necessarily be enhanced by restricting student mobility, it was argued. In any event, countries that had engaged in past hostilities with the United States were sending very few students to attend American colleges and universities. Afghanistan, for example, was sending about 20 students in 1999–2000; Iraq had sent 50; and Iran dispatched about 450 students to the United States. Zong Wa, managing editor of the International Education Expo, summed up the uncertainty felt by onlookers abroad as they tried to make sense of the fast-changing policy environment in the United States. Writing in the Beijing *Morning Press*, he observed, "We feel the September 11 attacks have exacerbated an already volatile and confusing situation. . . . "[16] What the months ahead had in store seemed impossible to predict. Many others echoed the same sentiment.

In the face of countervailing public opinion, NAFSA leaders dropped their opposition to the idea of a monitoring program within a matter of months after 9/11. Meanwhile, in the fall of 2001, Congress passed the USA Patriot Act, which reinstated a January 30, 2003, deadline for colleges and universities to begin using SEVIS. The following July (2002), the INS released SEVIS for collegiate use. Under its provisions, this $36.8 million system placed responsibility squarely on colleges to report whether their foreign students had in fact arrived on campus and were actually enrolled to take courses. In addition to making reports as they had to when students did something that affected their visa status (such as changing an academic major), colleges henceforth would be obliged to update students' files at the beginning of each term. If the INS did not receive an update, the student would be considered "out of status" and liable for penalties, including arrest and summary deportation.

Continuing Controversy Surrounding SEVIS

In the mid-2000s, college and university personnel were still scrambling to bring their respective institutions into compliance with the mandated database program. While many remained skeptical, defenders of the system were persuaded that the tracking system would identify the whereabouts of all foreign students and raise warnings if a foreigner entered the country on a student visa but never showed up for classes. But according to a 2003 survey conducted by the American Association of Collegiate Registrars and Admissions Officers, technical glitches were preventing full implementation of the system as schools tried to access information and use the database as intended. It was also being reported from many sources that the cost of placing the system in operation was far in excess of what the federal government had earlier predicted.

In February 2004, passions erupted on the campus of the University of Massachusetts at Amherst when more than 200 students refused to pay a new fee charged to the campus' 1,600 international students. The $65-per-semester fee, officials explained, would help make up for cuts in the budget of the school's International Programs Office. A small portion of the fee, it was further disclosed, was intended to help pay for the federally mandated monitoring system for students from other countries.

Critics alleged that the fee was wrong-headed on several counts. First and foremost, they argued, students should not be asked to pay for their own surveillance, and international students should not be asked to shoulder the whole burden of paying for international programs by themselves. "It's not ethical to charge someone for their own surveillance . . . no matter what percentage of the fee is for SEVIS," declared George Liu, a Chinese graduate student. "The fee-for-service model is also problematic, because you can't single out every group. You can't charge students of color for diversity programs, or women students for the women's center."[17]

Led by the Graduate Employee Organization, a campus union representing 1,000 international graduate students employed at the university, opponents of the fee staged a lively protest outside administrative offices and issued an on-line petition condemning the new charge. Yet as university administrators were quick to point out—SEVIS was mandatory. The school had no choice but to find funds to help pay for it. All that remained to resolve was where the monies would come from and who would pay—domestic students the system was supposed to protect, international students, or all students alike.

Despite continued criticism, by early 2007, the federal government's Student and Exchange Visitor Program (SEVP) had its database program more or less in full operation, now relatively free of the technical glitches that had plagued its introduction years before. SEVP's Web site described the agency and its mission: "It acts as the bridge for varied government organizations which have an interest in information on foreign students," it was explained. "SEVP uses web-based technology, the Student and Exchange Visitor Information System (SEVIS), to track and monitor schools and programs, students, exchange visitors and the dependents throughout the duration of approved participation within the U.S. education system." The objective, the narrative continued, was to provide information "so that only legitimate foreign students or exchange visitors gain entry to the United States."[18]

Spirited opposition from some quarters notwithstanding, SEVIS had evolved into a system of formidable proportions, absorbing information about foreign students attending many different types of educational institutions and programs throughout the United States. Although other data sources focused only on institutions of higher education, SEVIS' computations included secondary schools and vocational schools as well, plus data drawn from 13 categories of the Exchange Visitor Program: university or college student, high-school student, physician, au pair, camp counselor, summer work and travel, trainee, government visitor, international visitor, professor, research scholar, short-term scholar, specialist, and teacher. Aggregated data supplied by SEVIS in early 2007 cited a total of 8,771 approved schools, with a total of 10,519 campuses, 611,581 active students, 1,433 active exchange programs, and 154,471 active exchange visitors—all falling within the SEVIS purview.[19]

Homeland Security and International Students: The Ongoing Debate

In the immediate aftermath of the World Trade Center and Pentagon attacks and the popular outrage they prompted, it was widely assumed that large numbers of foreign students would elect not to pursue their studies in the United States. And for several months, those predictions seemed borne out. But as one commentator observed in September 2002, exactly a year after the bombings, " . . . Terrorist attacks may have left their imprint on the American psyche, but they do not appear to have dimmed the attraction of the United States for foreign students. . . . The number of students who are coming from the Middle East, some of whom say they are fearful of being attacked because of their religion or ethnicity, actually increased 4.6 percent last year over the previous academic year, although that rise masked drops among students from Oman, the United Arab Emirates, and a few other countries." The editorialist concluded, "For reasons that are not clear, the number of students coming to the United States from Western Europe fell by 1.7 percent. But growth in other regions, such as Asia, Central America, and Africa, more than offset that small decline."[20]

Indicative of the strained climate foreign students might encounter—a factor assumed to influence students' decisions to study in the United States or elsewhere—was a White House announcement in April 2002 that President George Bush was considering barring international students from studying in certain academic fields, including areas that had a "direct application to the development and use of weapons of mass destruction."[21] Administrative officials, it was disclosed, had been discussing for some time whether there were "sensitive courses of study" to which foreigners should not be admitted—chemical engineering, nuclear technology, biotechnology, advanced computer technology, and robotics, among them. Jim Griffin, assistant director for social and behavioral sciences at the White House science-policy office, was quoted as explaining, "We're working very hard to balance the needs of enhancing homeland security with our recognition of the valuable contributions that international students make to our academic communities."

College officials, it was reported, had responded with a request that the administration include them in discussions. "We understand that they are talking about national-security issues and that some of these discussions have to take place in secret," declared George Leventhal, a policy analyst with the Association of American Universities. "But if the government is going to issue a new policy that affects universities, clearly universities want to be consulted first."[22]

College and university administrators representing the Association of American Universities, the American Council on Education, and the National Association of State Universities and Land-Grant Colleges went further. "We believe," they were quoted as saying, "that the most fruitful approach will be to concentrate on preventing initial entry into the U.S. of individuals who pose a potential security threat." They continued, "The fundamentally open character of our higher-education system makes it [extremely difficult], and probably impossible, for colleges to restrict students already present in the country from gaining access to information that is made available to other students."[23]

Invoking a now-familiar argument, critics of the proposed curricular restrictions questioned anew whether it was really in the national interest to shut foreign students out of certain courses or academic programs. "Suppose we had not let [U.N. Secretary General] Kofi Annan or the king of Jordan study in this country," came the rhetorical question. "How would they feel about us now?"

The roster of world leaders who had studied in the United States but who might have been turned away under more restrictive circumstances was a lengthy one: Afghanistan President Hamid Karzai, Mexican President Vincente Fox, British Prime Minister Tony Blair, French President Jacques Chirac, South African President F. W. DeClerk, Israeli Prime Minister Ehud Barak, Argentinean President Paul Ricardo, Bolivian President Sanchez de Lozada Bustamante, South Korean Prime Minister Kang Young Hoon, German Prime Minister Ernst Carl Julius Albrecht, and countless others. As foreign students in America, each of these leaders—and many more—had learned about the nation and its people at first hand, an exposure that allegedly could not help but benefit the country that had hosted their stays.

Ira Pohl, a professor of computer since at Santa Cruz, took the point one step further. It would still be shortsighted, he asserted, for the government to apply its restrictions only to students from terrorist-supporting nations. "A lot of people who come from those places stay and flourish here, or go back and become moderating influences in their own countries," he alleged.[24] Much the same point was registered by Secretary of State Colin Powell in August 2002. "I see the benefits of international education every day," he declared. "In my daily activities I encounter world leaders . . . who participated in an exchange program or studied here or abroad. . . . People-to-people diplomacy, created through international education and exchanges, is critical to our national interests."[25]

Perhaps no more eloquent a defense of international exchanges was advanced in the early and mid-2000s—an era marked by so much turmoil, policy confusion, and paranoia—than that supplied by Victor Johnson, associate executive director for public policy at NAFSA: Association of International Educators. There exists, he claimed, a "residual affection and respect" for the American people held by foreigners, even when particular American policies are heartily detested. Those positive feelings derive from the American Idea which continues to inspire the world. And it is nurtured, Johnson asserted, "through exchanges, through programs that bring people into contact with America and Americans," including those that have brought millions of foreign students to the United States.[26] He concluded with an appeal:

> It is time for the United States to stop committing unilateral disarmament in the battle of ideas, values, and beliefs—especially in the Arab and Muslim world—that is central to success in the war on terrorism. International education is part of the solution to terrorism, not part of the problem. When, in the name of security, we carry controls on exchange so far that we threaten exchange itself, we don't increase our security, we decrease it, our leaders must step up and rescue one of our best hopes for creating the secure, peaceful world in which we all want to live.[27]

Visa Restrictions

Certainly the greatest obstacle international students faced by the thousands as they sought permission to study in the United States were new visa restrictions and stringent identity checks imposed in the wake of the September 11 attacks. As Daniel Walfish nicely phrased it, writing in the *Chronicle of Higher Education* in November 2002, "For many students trying to reach the United States, the past few months have amounted to endless waiting in a stalled security line."[28] Students seeking visas were reporting numerous hold-ups and a frustrating lack of information about the causes of the delays as the semester progressed past the point where applicants could enroll. Similarly, more than a few U.S. colleges and universities that had reserved research slots and teaching positions for foreign students were left with staff vacancies impossible to fill at the last minute.

Everywhere the story was the same. Sun Zhi, a graduate of China's prestigious Tsinghua University, who had been awarded a scholarship for graduate work from an American university, was turned down for a visa by the U.S. Embassy in Beijing. "It's all because of 9/11," he alleged angrily. "Had I applied one year earlier, I'd have easily gotten a visa." After having been rejected twice, he abandoned his dream of studying in the United States. "Many of my classmates have changed their plan to go to American universities to earn a Ph.D. due to the tightening of visa approvals," he was quoted as saying, "Since it has become so difficult, we think it's a waste of time to apply."[29] Clearly, the United States no longer seemed to be a friendly destination.

Sun Zhi's experience was replicated around the globe. "This is catastrophic," complained Irving Lerch, director of international affairs at the American Physical Society. "The government is foundering at the moment. It's casting about, trying to find a procedure that will work."[30] Everywhere student complaints spoke to the same problems: long lines, procedural confusion, uncertainties about how to fill out complex documents, ham-fisted enforcement of new policies, arbitrary or needlessly-rigid interpretations of the rules, endless delays, and unexplained rejection of applications—for thousands of applicants worldwide, the experience was sufficient to dissuade them from trying to come to study in the United States. Equally frustrating was the experience of many who left the country for brief visits home but then found the way barred when they sought to reenter the country to continue their studies.

"All the signs are that things look to be going from bad to worse," predicted Victor C. Johnson, commenting on the situation. "We are shooting ourselves in the foot if . . . students, who want to pursue higher education in the United States, can't get in or find it too much of a hassle to try."[31] No legitimate interest was served, least of all those of the United States, he added, by throwing up needless obstacles for foreign students who posed no security risk whatsoever.

The Enrollment Decline

Concerns over visa delays and rejections, it was widely predicted, would inevitably translate into a reduction in the total enrollment of foreign students nationwide. Others, however, felt that economic factors would come into play in the months to come, in particular a stagnant U.S. economy that would have the effect of shrinking the available pool of financial aid available to graduate students. Also, it was said, countries like China could be expected to invest more of their own resources in new graduate programs, making the idea of studying at home more appealing to their own students. Likewise in India, a major exporter of students to the United States, a booming economy might encourage recent college graduates to remain at home and enter the local workforce instead of pursuing advanced studies overseas.[32]

Early enrollment reports were fragmentary and conflicting. For the 2002–2003 academic year, the total number of foreign students enrolled showed an increase, but at a rate less than 1 percent compared with previous years when the norm was around 5 percent. The first of a series of surveys conducted by NAFSA: Asso- ciation of International Educators, the Association of American Universities (AAU), the National Association of State Universities and Land-Grant Colleges (NASULGC), the Institute of International Education (IIE), and the Council of Graduate Schools (CGS), released in March 2004, found that 47 percent of the 250 colleges and universities that responded received far fewer graduate applications for the fall term of 2004 than they had the year before. Among them were nineteen of the twenty-five U.S. research institutions that enrolled the largest number of international students. Commenting on the decline, Victor Johnson of NAFSA remarked, "The declining rates are clearly connected to a series of actions taken by the U.S. government to make it progressively more difficult for people to enter the country."[33]

Six months later, in September 2004, the Council of Graduate Schools released the text of a second survey, this time of 126 institutions, indicating an 18 percent decrease in foreign student admissions between the fall of 2004 and the previous fall. The largest enrollment drops reportedly were 34 percent from China, 19 percent from India, and 12 percent from South Korea, the three countries that had been sending the most students to the U.S. in recent years.[34] John Gravois, writing in the *Chronicle of Higher Education* for September 17, 2004, commenting on the survey, called it "a new round of sobering data on the continuing downward trend in foreign graduate admissions since the terrorist attacks of 2001."[35]

Additional data were forthcoming soon thereafter that seemed to confirm preliminary reports of an enrollment decline. "We don't know if this is the tail end of a short-term trend or part of a long slide," observed Peggy Blumenthal, vice president for educational services at the Institute of International Education. But the findings, she said, was a "wake-up call" that pointed to a need for urgent action. Despite moves to eliminate or at least reduce impediments for foreign students, especially visa delays, more must be done, she declared. A concerted effort was needed to counter perceptions in many countries that the U.S. no longer welcomed foreign students.[36]

On November 4, 2004, the Council of Graduate Schools (CGS) issued yet another report that contained a number of disquieting findings.[37] Among them was that with 68 percent of its 450-member graduate schools responding, a significant decline had taken place from the previous year in first-time international graduate enrollments. The drop translated into a 6 percent decrease overall. First-time enrollment from China reportedly had decreased 8 percent and from India, 4 percent, the two countries with the largest number of students ordinarily seeking admission to U.S. schools. Enrollment in fields of study usually filled to near-capacity by foreign nationals likewise had declined, marked by a 10 percent drop in life sciences and agriculture, business (a 12 percent decrease), and engineering (an 8 percent decline). These fields, significantly, had undergone the steepest declines, while physical sciences (up 6 percent) were the only fields of study sustaining an increase in first-time enrollment.

Eighty percent of the twenty-five institutions with the largest international student enrollments had responded to the survey. Collectively, they reported a 9 percent drop in first-time enrollments, compared with an overall 4 percent drop among schools outside the top twenty-five. The more pronounced decrease among the larger schools, the council pointed out, was owed to the fact that the top twenty-five schools polled also typically enrolled the largest number of students in engineering, a field of study where a large decline in enrollment was apparent.

The council identified a 28 percent decrease in international graduate applications and an 18 percent decline in international graduate admissions.[38] The main reason that the 18 percent decline in admissions did not translate into as large a decrease in first-time enrollment was a substantial *increase* in admissions. In 2003, for example, 38 percent of admitted international graduate students enrolled in American graduate schools, while in the year 2004 that figure stood at 43 percent.

From its annual survey of graduate enrollment and degrees for 2003, the CGS found that after an 8 percent decline the previous year, first-time international graduate enrollment had gone down 10 percent between the fall of 2002 and the fall of 2003. Public universities and research-intensive schools showed the largest decline (12 percent for each), while the decline in international enrollment was somewhat less for private institutions. Nonetheless, across the board, there were 6 percent fewer first-time international graduate students enrolled than in 2003.

November 2004 also marked the release of findings from a separate survey conducted the month preceding of foreign student enrollment data from 480 contributing institutions.[39] Thirty-eight percent of 431 institutions were reporting a decline in *new* international *undergraduate* enrollments for fall 2004, as compared with the total for fall 2003.

Of the responding doctoral/research institutions that each enrolled more than 2,500 international students (twenty-eight universities), 39 percent reported declines of varying magnitude. A total of 44 percent of 404 institutions queried reported a decline in *continuing* international *undergraduate* student enrollment. More than half—54 percent— of the twenty-eight doctoral/research schools similarly acknowledged declines. For undergraduates, more institutions reported decreases than increases in the number of students from select countries, including India, Indonesia, Pakistan, and Saudi Arabia.

So far as *new* international *graduate* student enrollments were concerned, 74 percent of 253 institutions once again reported declines. Among research universities, 56 percent had experienced substantial losses. The general trend reported was higher rates of decline among larger institutions. Approximately one-third of all colleges and universities reporting indicated fewer Chinese graduate students compared to 23 percent reporting increases; for India, the numbers were 36 percent claiming drops and 23 percent reporting an enrollment increase.

Approximately three out of five research universities reported a decrease in graduate school *applications* by international students, with some of the largest experiencing a decline of 30 percent or more. When asked to define the top factor contributing to the reported enrollment decrease, 47 percent simply noted a declining number of applications received.[40] Twenty-nine percent identified visa troubles (both delays in issuance and denials) as the major factor responsible.

Equally troubling were findings from the 2005 International Graduate Admissions Survey I.[41] Consistent with previous trends, it documented further declines. Applications for study in the U.S. had dropped 5 percent overall, following a 28 percent decrease from the year before. Whereas the magnitude of the decline was not as dramatic compared with the preceding year, the volume of applications for 2005 had not returned to 2003 levels, and the basic direction remained negative.

Leading losses were a 13 percent drop in applications from China, followed by a decrease of 9 percent from India, and 6 percent fewer applications submitted by prospective students from the Middle East. The two most common fields for international students—engineering and business—were marked by declines of 7 percent and 8 percent, respectively. With one exception (humanities), fields such as education, life sciences, physical sciences, and social sciences likewise showed drops ranging from 1 percent to 4 percent for the 2005–2006 school year. During the preceding two years, the same fields of study had all registered significant application declines ranging between 20 and 24 percent.

Responses to the Enrollment Decrease

Debra W. Stewart, President of the Council of Graduate Schools, spoke for many in identifying the main reasons for the decreases cited in successive studies. "The three primary factors leading to declines in international graduate applications, admits, and enrollment," she declared, "are increased global competition, changing visa policies, and diminished perceptions of the U.S. abroad."[42] Nils Hasselmo, President of the Association of American Universities, offered much the same assessment. "The major factors," he observed, "are U.S. visa policy, increased international competition, and perceptions that the United States is no longer a welcoming country." He added, with understatement, "Each of these areas needs attention."[43]

Marlene M. Johnson, executive director and CEO of NAFSA, warned that if the decline in applications from international students continued at prevailing rates, even greater losses of students might be expected in future. The good news, she commented, was that public officials allegedly were becoming aware of the seriousness of the problem and had begun addressing some of the obstacles that discouraged or prevented legitimate students from coming to the United States. The bad news, Johnson cautioned, was that if the U.S. were to lose the market for international students, it would be like "losing a forest to a fire. It happens very quickly. But winning back the market is like re-growing the forest. It takes time and effort."[44]

Philip Altbach of Boston College sounded a decidedly pessimistic note. The immediate future, he claimed, to him appeared "bleak" so far as attracting more international students to American colleges and universities was concerned. The United States, he suggested, had arrived at a "tipping point," a place marking the real beginning of a decline in the nation's international academic standng.[45] That former prominence, he predicted, would not easily be regained in any immediate or foreseeable future. On the contrary, in his opinion the situation was likely to worsen even if the U.S. launched a vigorous initiative to regain its former "market share" of students. CGS President Stewart, on the other hand, offered an upbeat perspective. "Declines," she remarked, commenting on findings from the several polls conducted in 2004, "are not nearly as great as some had feared. It is encouraging that the graduate schools are battling declines by streamlining the admissions processes, enhancing their use of technology, and forming international partnerships."[46]

Reactions to the apparent enrollment decline from the media came swiftly and continued throughout the first three-quarters of the decade of the 2000s. An editorialist for the New York *Times* (November 29, 2004) was among several pundits who pointed out that educating foreign students in the U.S. had grown to the point where it amounted currently to a $13 *billion* industry, and therefore that anything that adversely affected international enrollments needed to be taken very seriously.[47] More important still, ran the now-familiar refrain, was the hope and expectation that when they returned home, international students destined for high office in their respective countries might carry back with them a stronger commitment to American ideals and democratic values.

The theme of cultural exchange as a two-way street was sounded repeatedly. "We need them [international students] on campus," opined one student. "I'd hate to see us ever not having them—not for the money, but for how they enhance our campus. It's better for us to have foreign students . . . take our ideas back home. Closed doors could breed more terrorism."[48]

J. Michael Adams, president of Fairleigh Dickinson University, did not mince words in delivering his take on the problem. We should be "very worried" about the loss of students, he declared. The drop in numbers, he insisted, though small percentage-wise, could be accounted for in large part by a growing perception abroad that America was plagued by a slumping economy, a plethora of burdensome bureaucratic regulations, and "a rising tide of arrogance and intolerance." Adams cited various surveys, including a major poll conducted by the Pew Research Center, indicating that anti-American views had become pervasive throughout the Middle East and were intensifying in Europe as well. Quoting from *Business for Diplomatic Action*, Adams warned, "We're losing friends at an alarming rate and more people are boycotting American products." Lack of confidence in America, even open hostility, was undoubtedly contributing to the loss of international students, he claimed. Increasingly, those seeking an English-language study abroad experience, he pointed out, were turning away from the United States in favor of Canada, Australia, and Great Britain.[49]

NAFSA's Victor Johnson reiterated the many ways in which the presence of foreign students on campus helped enrich the collegiate experience. Internationals, he pointed out, added needed diversity to the student body and provided many American students with their first real contact with people from other cultures. Also, as many others had pointed out previously, he took note of the fact that upward of 70 percent of all foreign students were paying their own way out of their own pockets and hence contributing to the financial well-being of the schools they were attending.

Johnson likewise invoked the often-reiterated point cited previously, that it was not unrealistic to assume that whatever favorable impressions of America and Americans international students might acquire while in residence would accompany them and influence others on their return home.[50] Arguably, those residual feelings of goodwill, shared with friends and associates, might help on a long-range basis to foster more positive attitudes toward the U.S. and its people.

Finally, Johnson noted, as had many others before him, that international students typically filled under-enrolled science courses that colleges might otherwise find difficult to offer; and they helped staff such classes as teaching and research assistants, especially in the hard sciences and applied technology fields. In the same vein, Fareed Zakaria, editor of *Newsweek International*, felt that the single "most deadly" effect of declining international enrollments was "the erosion of American capacity in science and technology—research done by foreign students." The U.S. has never produced enough homegrown scientists and engineers to meet the nation's needs, he claimed. The shortfall has always been made up by foreign students who choose to remain and take up permanent residence in the United States. International students, he further pointed out, made up more than half of all students enrolled in scientific and engineering courses of study. "The dirty little secret about America's scientific edge," Zakaria commented, was that "it's largely produced by foreigners and immigrants. Americans don't do science anymore."[51]

Practically everyone agreed that U.S. visa regulations and procedures were a major factor behind the decline in international student enrollments.[52] An editorial in the Daytona Beach *News-Journal* on November 26, 2004, bemoaned the "mishmash of improvised, inconsistent rules" put in place, regulations whose practical if unintended effect was to dissuade prospective students from applying to come to the United States. "We'll never know if a potential terrorist or two were kept out of the country among the tens of thousands of foreign students choosing to go elsewhere," the editorialist continued. "We do know that the 'unilateral disarmament' in brains is hurting research, impoverishing campus culture and projecting the wrong image about America. For a nation of immigrants, that's an unexpected, remediable disgrace."[53] Wrote a columnist for the Indianapolis *Star*, "Considering that education is one of the nation's ... most valuable commodities, the U.S. needs a more flexible system of dealing with foreign students."[54]

Anecdotal evidence was overwhelming, declared the Cincinnati *Post* (November 30, 2004) that would-be students were finding post-9/11 visa procedures "cumbersome, time consuming, demeaning and seemingly designed to keep foreigners from visiting the United States."[55] Applicants were now being made to run a gauntlet, extending from numerous required consular visits and intimidating interrogations at the front end all the way through endless red tape to an uncertain reception at the visa holder's final port of entry into America. Surely, it was said, this was a "fixable problem."

It was not simply a matter of safeguarding national security versus granting immigrants access or—as the *New York Times* put it—excluding a dangerous few while excluding the helpful many. Visa policies needed to be kept rational, even while responsive to legitimate security considerations. "Consular officials know that they face career-threatening punishment if they are too lax, but face little sanction if they are too strict," it was observed. Hence, their understandable inclination allegedly was to err on the side of caution and deny more visa applications than they approved. These "perverse incentives" for bureaucratic inefficiency had to be eliminated as promptly as possible.[56]

Making much the same point, the *Detroit Free Press* for November 23, 2004, declared, "At a time when America needs all the friends it can get, the country is missing out on one of its best opportunities to cultivate good will and an appreciation for the democratic way of life."[57] Warned David Dinkins, former mayor of New York City, "If these trends continue—with foreign students being shut out of American schools ... the United States runs the risk of being shut out as [a] competitor in the global marketplace of commerce, trade, diplomacy and the healthy exchange of ideas."[58]

Enrollment Stabilization

By the middle of the century's first decade, the enrollment decline seemed to have leveled off somewhat compared to the two or three preceding years. Contrary to earlier dire predictions, no dramatic or precipitous drop in foreign student applications or admissions for the 2005–2006 academic year had materialized, thanks at least in part to fine-tuning of visa processing and immigration procedures. Another causative factor at work might have been—justifiably or wrongly—the fading of the public's fears about terrorism and therefore lessened concern about security issues. Downward trends evident over the years from 2002 on were not entirely reversed. But neither had they worsened appreciably.

Preliminary figures from the start of the 2005–2006 academic year served to calm earlier fears that the foreign student enrollment would continue on a downward track. In actual fact, few changes from the 2004–2005 were registered. As in earlier years, India remained a major exporter of students to the United States, though not quite at the level typical of the early 2000s. Early figures for China suggested a slight decline as well, while Japan, Turkey, and Pakistan remained steady.

Enrollments by fields of study exhibited no major changes from patterns of the immediate past. Leading choices as before were engineering, mathematics, computer sciences, and business, although slightly under the level of demand common earlier in the decade. Schools hosting the largest number of international students were the familiar ones: University of Southern California, Columbia, Purdue, New York University, and the University of Texas at Austin. States with the biggest foreign student aggregates included California, New York, Texas, Massachusetts, Florida, Illinois, Michigan, Ohio, Pennsylvania, and Indiana. Superficially, much had remained the same.

International Competition for Students

What had changed, however, was as noted previously a significant increase in competition for foreign students from such other English-speaking countries as Australia, New Zealand, Canada, and Great Britain. These nations—in common with the United States—had once been content to sit back passively and await whatever foreign enrollment applications came in. Now, belatedly realizing the cultural and educational advantages of attracting foreign students to their campuses (not to mention the considerable financial benefits), they had begun to mount vigorous recruitment campaigns, sometimes complete with financial aid, one-stop information and admission offices, scholarship packages, and promises of special mentoring and support, along with other inducements. And to the list of European competitors vying for students, increasingly could be added such Asian and South Pacific nations as South Korea, Malaysia, and Singapore.

From the 1990s on, virtually all member nations of the European Union were beginning to address the challenge of coordinating their respective student recruitment efforts, enrollment procedures, transfer policies, and academic transcript reporting. The aim was to enhance the mobility of international students and, possibly, to attract more of them to apply within the European Union sphere. All the while, so-called "feeder" countries, such as India and China, had begun offering incentives to keep more of their students at home rather than having them venture abroad for their education.

It remained more or less true to claim that in the opening years of the twenty-first century the United States was by far still the most popular choice by far for students seeking higher education abroad. What was not always apparent behind the reports of increasing enrollments was the arduous effort it required to induce them to apply. "For many people, both inside academe and out," observed Beth McCurtrie, writing in the *Chronicle of Higher Education*, "the assumption is that American colleges don't need to work to attract foreign students." That might be true for a handful of elite institutions as well as "those willing to accept any warm body that walks in the door," she allowed. But for hundreds of others, "those solid but not stellar universities whose names are likely to evoke blank stares when mentioned abroad—each application is usually the result of a lot of hard work."[59]

As U.S. colleges and universities came around to the view that international students did in fact represent valuable assets—a "stock of intellectual capital"—for their campuses, not only did competition for foreign enrollments stiffen between the United States and other nations, but it began to create rivalry between American institutions of higher learning as well. Once the province of small private liberal-art colleges, now larger universities began mounting multi-country promotional tours, fairs and exhibits, all intended to interest students in attending the particular sponsoring institution.

Interestingly, major beneficiaries of the competitive marketplace were American community colleges. Touting the advantages of being able to offer more individualized attention, close mentoring and career counseling, smaller classes, and less costly tuition, two-year community colleges began registering significant enrollment gains. In the period from the 1993–1994 academic year through 2002–2003, for example, the number of foreign registrants at 1,200 two-year colleges swelled by 60 percent, from slightly more than 60,000 students to nearly 97,000. That figure was double the overall growth rate of the number of international students at all postsecondary colleges and universities within the same time span.[60]

The Bologna Process

In May 1998, the ministers in charge of higher education in France, Italy, the United Kingdom, and Germany met in Paris at the Sorbonne to sign a so-called "Sorbonne Declaration." Basically, it called for the wholesale "harmonization" of the architecture of Europe's diverse higher education systems. On June 19, 1999, twenty-nine European ministers (later joining in were representatives from Cyprus, Turkey, Croatia, Liechtenstein, Albania, Andorra, Bosnia, Herzegovina, Serbia, Montenegro, Russia, and the Yugoslav of Macedonia) convened in Bologna to establish a common "European Higher Education Area" (EHEA) within which drastic and far-reaching educational changes were planned. Nowhere had an educational reform movement on such a grand scale been attempted before.

In the Bologna Declaration, signatories pledged themselves to working toward a system of easily readable and comparable degrees (bachelor's, master's, and doctoral), an institutional structure consisting of two separate and distinct layers or levels (undergraduate and graduate); and a common system of academic credits. Several other quality assurance measures were promised as well.

Two years later, in May 2001, officials reunited in Prague, Czech Republic. Earlier goals were reaffirmed and new ones added, most notably an official intent to enhance the "attractiveness" and "competitiveness" of European postsecondary education to other parts of the world (thereby attracting foreign students to the European higher education area and facilitating transfers within member states). Lloyd Armstrong Jr., provost emeritus of the University of Southern California, characterized the development as "the most fundamental re-envisioning of higher education undertaken as a consequence of extra-national considerations."[61] Robert Sedgwick, editor of *World Education News and Review*, opined, "The reforms . . . will no doubt impact Europe in many ways, but they also hold significant implications for international educational exchanges in the United States."62

Sedgwick went on to comment:

> In the last 10 years, the market for international students (particularly students from Asia) has heated up considerably, and Europeans fear that if they don't act decisively they could end up falling behind while the United States and its competitors (Australia, Canada and the United Kingdom) corner the market.... Newly reformed higher education in the European Higher Education Area, along with relatively low tuition fees, will hopefully make Europe a viable option for many international students who cannot or will not pay the high cost of education in North America or Australia.... In the future, with its transparent and flexible higher education system, Europe will offer an attractive alternative. The EHEA may end up challenging American dominance in international higher education, in much the same way that the European Union has become a counterweight in international trade vis-à-vis the U.S. and Japan.63

Philip Altbach of Boston College, writing in *Connection: New England's Journal of Higher Education*, offered a trenchant analysis of the transformation European higher education was expected to undergo under the terms of the Bologna process. "Significant changes underway in higher education worldwide will inevitably affect [America's] international role," he declared. The Bologna Agreement, as he interpreted it, was a "wild card" making it difficult to frame estimations of what the future was apt to bring. "What Bologna will mean for European students and staff coming to the United States is not yet clear," he allowed. "Another question is whether harmonization will build barriers to students from outside Europe—Asians, Africans and Latin Americans as well as those from the United States." Summarizing some of the changes pending on the international scene, he concluded, "How ... changes will affect the United States ... remains somewhat murky, but it is clear that global competition will increase as new providers of international education services enter an expanding market."64

Looking Backward and Ahead

Early in the spring of 2006, Hey-Kyung Koh Chin, a senior program officer of research and evaluation at the Institute of International Education and editor of the *Open Doors Report on International*

Educational Exchange, provided a helpful summary of trends and shifts apparent in comparing data for the academic year 2004–2005 with the year preceding.[65] (In one sense the comparisons were of interest as much for the *lack* of changes shown as for real differences transpiring over time.) Regions of origins of international students, she noted, had remained virtually unchanged in 2004–2005. The largest proportion of international students came from Asia (57 percent), up 1 percent from 2003–2004. Enrollments from Europe (13 percent) and Latin America (12 percent) remained substantially unchanged, while enrollments from Africa (6 percent) declined slightly. Those from North America (5 percent) and the Middle East (6 percent) remained virtually the same.

The distribution of international students by state also had changed very little and remained heavily geographically concentrated (mostly in thirteen of the states). The five leading host states as of the fall of 2004 were California, New York, Texas, Massachusetts, and Florida. New York City was once again the biggest metropolitan area hosting foreign students, followed by Los Angeles; Boston; Washington, D.C.; and Chicago. The leading twenty-five receiving institutions were all doctoral research institutions, each of which hosted 3,000 or more international students.

Consistent with earlier trends, it was found that international students were studying business and management most frequently, followed by engineering. Noteworthy was a finding that there had been a 25 percent decline in foreign student enrollments in computer sciences and mathematics. Other fields of study showing the largest enrollment decreases included social sciences (15 percent) and fine and applied arts (12 percent). Increases were registered in the physical and life sciences (11 percent) and intensive English language (8 percent).

As had been found true in previous years, approximately two-thirds of all international students in the United States were dependent on family and personal funds to support their studies. Also of interest was the finding that the gender ratio was becoming more balanced, with 44 percent female students in 2004–2005 versus 30 percent during the 1970s. Unchanged over time was the marital status of international students: 85 percent single.

With respect to total enrollments, Chin noted a "slowing decline" of 1.3 percent from the academic year 2003–2004, which followed a 2.4 percent decline the previous year, amid minimal growth of 0.6 percent in the year prior. These shifts followed in the wake of two or more consecutive years of 6.4 percent increases and nearly thirty years of healthy growth before that. In 2005, there were only 273 fewer foreign students enrolled than the enrollment total for the previous year, a point often overlooked or ignored in discussions about declining enrollments.

As of 2005, according to revised figures, 564,766 foreign students were enrolled in American postsecondary schools. And as in previous years, popular opinion in some quarters attributed the decline in total enrollment to tightened student visa review procedures put in place after September 11, 2001. Nevertheless, while its efforts were not always adequately acknowledged, the U.S. government actually had acted fairly promptly to add several hundred consular officers to handle via application backlogs, reduce waiting times, and increase the via approval rate to pre-9/11 levels in most countries. By and large they had been successful. Unfortunately, it could be argued, student perceptions lagged behind reality, thereby encouraging would-be scholars to pursue their studies in countries having what they imagined to be quicker visa-issuance procedures. The challenge was to counter an often-expressed perception: "Well, if it's that difficult to get into the U.S., then they can't be very welcoming once you arrive, right?"[66]

In early September 2006, the Council of Graduate Schools released the findings from its most recent poll.[67] The number of foreign students admitted to American graduate schools had increased in 2006 for the second consecutive year. A rise in applications, it was reported, would likely translate into a significant growth in graduate foreign enrollments later in the year. Meanwhile, admissions for the 2006–2007 academic year were reportedly up in most fields of study, with engineering seeing the biggest increase, up 26 percent over 2005–2006. Business applications likewise were up 12 percent. Many more students were admitted also from China and India, up 28 percent over the previous year for the former; up 20 percent from the latter. On the whole, there appeared to be warrant for cautious optimism about future international student enrollments in the United States.

Ellen D. Cohen, an associate director of Columbia University's International Students and Scholars Office, conceded there had been "terrible visa problems" for several years after the 2001 terrorist attacks. Notwithstanding, she emphasized, the situation had improved markedly by the fall of 2006, such that students attracted to strong research universities were choosing to look at the United States once again. A joint statement by the Institute of International Education in cooperation with several other professional associations the same year

Table 8.1 International Student Enrollments

Year of Enrollment	International Enrollment	Annual % Change	International % Total
1999–2000	514,723	4.8	3.8
2000–2001	547,867	6.4	3.9
2001–2002	583,996	6.4	3.9
2002–2003	586,323	0.6	4.6
2003–2004	572,509	−2.4	4.3
2004–2005	565,039	−1.3	4.0
2005–2006	564,766	−0.1	3.9

Source: Open Doors, 2006

Table 8.2 Countries of Origin of International Students in the United States, 2005–2006

Country	Student Total	Approx. Annual % Change
India	76,503	−0.05
China	62,582	0.01
Korea	58,847	0.10
Japan	38,712	−0.09
Canada	28,202	0.01
Taiwan	27,876	0.08
Mexico	13,931	0.07
Turkey	11,622	0.02
Germany	8,829	0.02
Thailand	8,829	0.02
United Kingdom	8,274	0.01

Source: Open Doors, 2006

Table 8.3 Highest International Student Enrollments by Host States, 2006

State	Enrollment	Annual % Change
California	75,385	0.01
New York	64,283	0.04
Texas	46,869	−0.01
Massachusetts	28,007	0.01
Florida	26,058	−0.01

Source: Open Doors, 2006

Table 8.4 Leading Host Universities for International Students in the United States, 2006

University	Enrollment
University of Southern California	6,681
Columbia University	5,575
Purdue University	5.540
New York University	5,502
University of Texas–Austin	5,395
University of Illinois	4,904
University of Michigan	4.649
Boston University	4,542
Ohio State University	4,476
State University of New York–Buffalo	4,072

Source: Open Doors, 2006

Table 8.5 Leading Fields of Study of International Students in the United States, 2006

Field of Study	Percent of Total	Enrollment
Business & Management	17.9	100,881
Engineering	15.7	88,460
Physical Sciences	8.9	50,168
Social Sciences	8.2	46,132
Mathematics, Computer Science	8.1	45,518
Fine and Applied Arts	5.2	29,509
Health Professions	4.8	27,124
Education	2.9	16,546
Humanities	2.9	16,480
Agriculture	1.4	7,883

Source: Open Doors, 2006

sounded much the same optimistic theme. Its findings, it was announced, while not comprehensive, suggested that the declines from previous years "may have worked their way through the system" and that overall enrollments could be expected to rebound within the near future."[68]

In January 2006, the U.S. State Department invited 120 college leaders to Washington to consider how to make American higher education more engaged with the world and to counter the perception that the United States no longer welcomed foreign students.[69] Secretary

Condoleezza Rice, one of several high-ranking officials to address the conference, told attendees that U.S. security depended on expanding the international exchange of students and scholars. She pledged to reduce or eliminate visa application problems, to increase the number of foreign students accepted for study in the United States, and, generally, to "make America more open while still maintaining [national] security."[70]

Karen Hughes, Under Secretary of State for Public Diplomacy and Public Affairs, in her address to the assembly, conceded that worldwide competition for hosting international students had greatly intensified in recent years. "Not too many years ago," she observed, "the United States was not only the best place to go, it was also one of the only places to go for serious study at the university level." She continued, "Today, hundreds of thousands of international students have many more opportunities to study at home, at centers of academic achievement in their own countries. . . . " Noting that many countries had begun competing aggressively to attract international students to their shores, she urged her listeners to help find "new and more effective ways to market the depth and diversity of American education overseas." Her challenge to conference participants, she emphasized, was to promote the United States "as the world's education destination through more active and effective marketing of United States higher education abroad."[71]

It remained to be seen whether the federal government in partnership with the nation's academic institutions would muster sufficient resources and energy to meet that all-important challenge.

"The long held model of the university as an ivory tower divorced from the everyday concerns of society is not tenable in a global village. A greater sense of involvement is imperative."
Bill Tierney, "Compass," *The Navigator* 6 (Fall 2006): 2

"America's colleges and universities have begun to see positive results from their . . . efforts to recruit international students and make them feel welcome."
Burton Bulag, "Enrollment of Foreign Students Holds Steady,"
Chronicle of Higher Education 43 (November 13, 2006): 9

"Foreign students provide links from our culture to other cultures. It is through them that many of our international trade partnerships are born. Foreign students who come to study here

go back to their countries as unofficial 'ambassadors' for America, as they share their experience in the U.S. with friends, family, and colleagues abroad."
American Immigration Law Foundation, *Immigration Policy Focus* (February 2003)

"The United States has been a destination for education and research for generations of foreign students and scholars, and this remains true today."
Jeanne Batalova, Migration Policy Institute (January 1, 2007)

Epilogue: Challenges and Imperatives

Security Questions in the Aftermath of September 11, 2001

Questions surrounding the issuance of visas for students seeking entry into the United States reflected one of the major security controversies that dominated discussions of American foreign policy throughout the early 2000s. The advisability of building what some critics disparaged as an "Orwellian" computerized tracking program (the system that became SEVIS) to monitor foreign students during their stay in the U.S. was another. Third, the proposal to deny international students access to certain "sensitive" courses of study and information deemed critical to homeland security represented still another issue prompting sharp divisions of opinion.

Possibly the liveliest debate of the period revolved around a series of critiques that threw into question the entire rationale underlying foreign exchange. The opening salvo came with the publication of a report entitled *Evaluation of the Foreign Student Program* by Harvard economist George Borjas, released in June 2002 under the imprimatur of the Center for Immigration Studies (CIS).[1] (An excerpted version entitled "Rethinking Foreign Students" had appeared two or three weeks before as an article in the *National Review*.[2]) Shortly thereafter, the CIS convened a panel on which Borjas was joined by the American Council on Education's Terry Hartle and other academic luminaries to discuss his broad-based attack on international exchange generally and the importation of foreign students in particular.[3]

Borjas's *National Review* paper opened with the observation that by the year 2000 the State Department was issuing more than 315,000 student visas for foreigners annually. But the "program," as he termed it, was now "so riddled with corruption and so ineptly run" that the Immigration and Naturalization Service (INS; U.S. Citizenship and Immigration Services) simply did not know how many foreign students were in the country or where they were enrolled. "There is little doubt," he alleged, "that the foreign student program has been spinning out of control for years.... There are few checks and balances to keep the number of foreign students at a manageable level, or to prevent foreigners from using the many loopholes to enter the country for reasons other than the pursuit of education." He continued, "Perhaps most important, the program has grown explosively without anyone asking the most basic questions: Is such a large-scale foreign-student program in our interests? What does it cost us? And what does it buy us?"[4]

Borjas laid out his argument. A foreigner who wants to study in the United States, he noted, begins by applying for admission to an educational or vocational institution. In order to qualify for a student visa, he or she must be accepted by an INS-approved institution. Once admitted, the school supplies a form attesting to the holder's approval for admission. The student takes this affidavit to a local consulate. He or she is interviewed and a decision is then made whether to grant the applicant a visa. Because a student may be accepted by several U.S. schools, the applicant likely ends up with multiple admissibility certificates. Inevitably, some end up on a black market where they are purchased at a high price to support illegitimate visa applications.

Once a student is issued a visa and enters the United States, there is practically no monitoring from then on, Borjas claimed.[5] Many entrants later adjust their immigration status and obtain a "green card" or permanent-residence visa. Lax monitoring, however, encourages many to stay in the country illegally, provided they are willing to be not too punctilious about their legal status. Some were—or are—bona fide students. Others, Borjas alleged, may have dropped out—but without any adverse consequences unless they happened to have come to the attention of the INS or some other investigative agency.

Obtaining admission from an INS-approved institution posed no serious obstacle, Borjas explained. He reported there were around 73,000 schools certified to mail admission eligibility forms, including no fewer than 400 in the San Diego area alone, including the Avance Beauty College, the College of English Language (new classes beginning every week), the Asian American Acupuncture University, and the San Diego Golf Academy. Other INS-approved "visas-for-sale" storefronts cited included the Nash Academy of Animal Arts (a pet-grooming school) in New Jersey, and the American Nanny College in Los Angeles. Borjas's verdict: "Anyone with the money can buy a student visa to enter the US." America, he alleged, had "effectively delegated the task of selecting immigrants to thousands of privately-run entities whose incentives need not coincide with the national interest."

By the same token, he suggested, neither do the financial incentives of large research universities necessarily align well with national security either. Academic institutions need low-cost workers to staff laboratories, and they depend on minimal-wage graduate teaching assistants to instruct large undergraduate classes. Foreign students, Borjas claimed, provide nearly an unlimited supply of willing workers who have not undergone any meaningful screening or background check.

Meanwhile, he complained, corruption abroad was rampant, most notably illustrated by a thriving industry of consulting firms that grease the wheels of the process, supplying bogus letters of recommendation, faked evidence of economic self sufficiency—and even professional actors to stand in for an applicant during the consular interview! "In short," he avowed, "the INS relegates the vetting of prospective students to an amazingly large number of institutions that benefit financially from the presence of foreign students, and to foreign consultants who brazenly misuse, distort, and pervert the system."

Borjas dismissed panegyrics celebrating the supposed benefits of hosting foreign students as nothing better than unsubstantiated rhetoric. If so, many benefits flowed from having American students exposed to foreign students, he suggested, foreigners for their part would be offering thousands of dollars to induce Americans to attend *their* universities, in order to facilitate *their* students' exposure to American visitors as well.

Nor was Borjas impressed with the often-heard claim that the United States benefits by skimming the best talent from other countries. In fact, he argued, there was no warrant for assuming that those students who remained in America were necessarily the most talented. A host of considerations (including marriage to U.S. nationals) might better explain who remained. Methods foreigners use to obtain student visas, and the ones American institutions use to recruit them, he suggested, "do not boost our confidence that only the best and the brightest show up on our doorstep."

Borjas concluded with a strongly worded indictment:

> It's not politically correct to say so, but the foreign student program may not be all that beneficial. Once we stop humming the Ode to Diversity that plays such a central role in the modern secular liturgy, we will recognize that the time has come for a fundamental reevaluation of the

program: Why should American taxpayers subsidize the tuition of the hundreds of thousands of foreign students enrolled in public universities? Is it sensible to give so many different institutions the authority to admit foreign students? Can we afford to ignore the national-security rationale for keeping some programs off-limits to students from particular countries? The remarkably powerful combination of INS ineptitude and the greed of the higher-education sector has perverted what seemed to be a sensible and noble effort into an economically dubious proposition and a national-security fiasco."[6]

Reactions to Borjas's allegation were swift and in some cases heated. At the June 2002 conference called by the Center for Immigration Studies, for instance, several participants and members of the audience clashed over the question about whether U.S. taxpayers were actually subsidizing the education of foreign students, as Borjas had alleged. Steven A. Camarota, director of research at the Center for Immigration Studies questioned whether state legislatures were aware of the extent to which subsidies did support foreign students. But Terry W. Hartle of the American Council on Education compared Camarota's comment to isolationist sentiments common in the nineteenth century. "The Yellow Peril is back," he reportedly charged.

"That's outrageous," Camarota responded, "You are basically a race-baiter."[7]

In a later forum, Hartle strongly defended the importation of foreign students to the United States. Not only did it serve the national interest, he insisted, but foreign collegians spending time in America was an economic bonus as well.[8] International students, he emphasized, did *not* receive special subsidies unavailable to their American counterparts. And insofar as they were spending billions of dollars on goods and services annually, the total of which far exceeded costs to the domestic economy, the financial advantage of hosting their time in America was substantial. (He also mentioned that according to the U.S. Commerce Department, higher education was the nation's fifth-largest service-sector export, ranking behind only travel, transportation, financial services and professional services.)

Every visitor to the United States, Hartle claimed, received a taxpayer subsidy if he or she landed at an airport, drove on a highway, rode a city bus, took a subway, visited a national park, went to a museum, or drank a glass of milk in a restaurant. Policymakers support the subsidy because the value of all the economic activity thereby generated far exceeds that of the subsidy. Expenditures by international students are no exception. Monies spent for tuition, rent, food, entertainment, clothing, books, and transportation all contribute to the economy and far exceed the value of whatever monetary support is forthcoming.

Citing a 1998 final report issued by the National Commission on the Cost of Higher Education, Hartle stressed the point that *no* student attending college in the U.S. pays the full cost of his or her education. Tuition charges in all cases are far exceeded by total costs. External revenue sources—alumni gifts, endowments, grants and contracts, auxiliary-service revenues, state monies—all subsidize tuition charged to students. As Victor Johnson, public policy director for NAFSA put it, the net economic benefit of foreign students living in the United States, even minus subsidies of all types, amounted to billions of dollars yearly.[9]

Allan E. Goodman, president of the Institute of International Education, hastened to counter some of Borjas's other claims. There is, in fact, no foreign student "program" as such, he insisted. Rather, hundreds of thousands of individual decisions are made by campuses and students, producing the half a million or more foreign students enrolled for credit at the nation's 3,800 accredited colleges and universities.[10] Moreover, Goodman added, academic institutions made their decisions in individual cases with reference to academic ability, not on the basis of security considerations. While acknowledging corruption and fraud in the system as a whole, Goodman pointed out that less than 2 percent of all those who had obtained visas to come to the U.S. were students at accredited schools—and they remained the most closely tracked group of all visa recipients.

"As we all work feverishly to reform the system of permitting and tracking entry into our country," he urged, "let us also work fervently to preserve the incalculable benefits of legitimate academic and cultural exchange that takes place in our nation's higher education system...." Goodman concluded his plea with a declaration. "Imagine a world in which our country does not protect the exchange of people and ideas across borders and does not promote academic cooperation and collaborative research among the best and brightest of all nations. It is just the kind where ignorance, hatred, and intolerance would flourish. And just what the terrorists are hoping still to create."[11]

The "Crowd-out" Effect

Borjas continued to serve as a sort of *bête noir* for internationalism advocates in the years following 9/11. His claims that the State Department had little control over the number and type of students being admitted to pursue their education in the United States was tacitly conceded to be accurate, at least until computerized tracking and monitoring became more widely available as SEVIS came online. Charges of overseas fraud were difficult to dismiss also, as were allegations that the process of issuing student visas by overseas consulates had grown positively sclerotic. The claim that the foreign student "program" should be conceptualized as a redistribution system, shifting wealth away from native workers and taxpayers and redistributing it to universities and foreigners, received little attention. The related question as to whether it was useful to think of the management of foreign students coming to America as the outcome of hundreds of thousands of individual decisions (as Goodman insisted) or as a discrete "program" (as Borjas termed it) received scant attention.

In an interesting succession of working papers, however, Borjas's next critique turned a spotlight on the effect of foreign student enrollments in certain graduate programs on U.S. nationals receiving degrees after completing the same academic course of study.[12] Although he found no evidence of a "crowd-out" effect for the typical American student, the impact of foreign students on native (U.S.) educational outcomes was found to differ dramatically across ethnic groups, and was especially adverse for white American males. Borjas's investigation did identify a strong negative correlation between increases in the number of foreign students enrolled in a specific program at a particular university and the number of white American men in the school's graduate program. The so-called "crowd-out" effect was found to be strongest at the most elite institutions studied.

A working paper issued in March 2006 extended the same general investigation.[13] It showed that a foreign-student influx into a particular doctoral field had an adverse effect on the earnings of doctorates in that field who had graduated at approximately the same time. A 10 percent immigration-induced increase in the supply of doctorates was found to lower the wage of competing workers approximately 3 to 4 percent. Borjas estimated that roughly half of the adverse wage effect could be attributed to the increased prevalence of low-paying postdoctoral appointments in fields that had softer labor market conditions because of large-scale immigration.

Finally, a third working paper examined the labor market impact of high-skill immigration into the science and engineering workforce.[14] The author concluded that the labor supply in these two specializations affected both earnings and employment opportunities. Increases in the number of foreign-born doctorates were found to have a significant negative effect on the earnings of competing workers, regardless of whether the competing workers were native-born Americans or foreign born.

Though the point was not made explicit, the obvious policy implication was that encouraging or even allowing large numbers of foreign-born PhDs in the sciences and engineering to remain in the country after completing their degree programs was apt to prove harmful for U.S. nationals competing for the same wages and positions.

The International Student Enrollment Decline Reconsidered

Estimates of the precise magnitude of the international-student enrollment decline that appeared to have transpired in the mid-2000s differed only slightly among interested observers. According to the Council of Graduate Schools, enrollment by foreign graduate students in the United States had declined by 8 percent between 2003 and 2004. The overall decline, as reported by the Institute of International Education, was 2.4 percent between 2002–2003 and 2003–2004. Figures reported by the Migration Information Service similarly indicated a drop in undergraduate enrollments between 2002 and 2003 of 0.4 percent and by 4.6 percent a year later.[15]

Enrollment of graduate students had initially risen within the same period, but had later declined by 3.6 percent by the academic year 2004–2005. The overall rate of decline had slowed by 2005–2006, lending credibility to the expectation that the enrollment drop would soon reverse and resume its upward climb as had been typical in earlier decades. Nevertheless, the picture remained decidedly mixed. America's share of all foreign students studying abroad, for example, had dropped from 25.3 percent in the year 2000 to 21.6 percent by the year 2004.

Furthermore, the situation was still somewhat unbalanced in terms of which nations were supplying the U.S. with visiting students. Just ten countries—India, China, South Korea, Canada, Taiwan, Mexico, Turkey, Germany, and Thailand—accounted for 59.5 percent of all foreign students enrolled in 2005–2006.[16] At the same time, the market share of other countries such as Australia, New Zealand, France, and Japan was increasing substantially.

Stuart Anderson, executive director of the National Foundation for American Policy, sounded the alarm. "Rather than retreating from our support for international student exchange—and foregoing its contribution to our national strength and well being—we must redouble our efforts to provide foreign student access to U.S. higher education," he declared.[17] Marlene M. Johnson, executive director and CEO of NAFSA: Association of International Educators, echoed Anderson's sense of urgency. The exact number of international students in the United States, she asserted—and whether the total was increasing or decreasing—mattered because it was a "surrogate for competitiveness," one of the measures employed to gauge whether the United States was succeeding in attracting talented people from around the world.[18] In this respect, she judged, the U.S. had not been doing well at all.

A 2006 position paper circulated by NAFSA expanded on the theme.[19] The United States was and is engaged in a global competition for international students and scholars, its authors affirmed. But it was not at all apparent whether the country was even aware of the competition. While other countries had been working hard to access the benefits gained from educating the next generation of world leaders and attracting the world's most talented elites, the United States, as they phrased it, seemed "curiously disengaged, content to compete with speeches, sound [bites], and photo ops." Most alarming, they continued, was that for the first time in memory the United States appeared to have lost its allure, that it seemed to be losing its long-held status as the destination of choice for international students.[20]

Today, the NAFSA Report continued, the collapse of U.S. competitiveness was apparent for all to see: the result of the transformation of the international student market, a clumsy and insensitive application of post 9/11 security measures, an increasingly negative American image abroad, and the virtual absence of a national strategy to address these problems.[21]

Reforming Immigration Law and the Visa System

Virtually all proponents of educational exchange were in general agreement about what was needed to restore America's competitiveness in the international arena, assuming the necessary political will to meet the challenge could be summoned. First, it was said, a through review and overhaul of immigration policies was required, the goal being the elimination of inconsistencies and illogical or counterproductive rules that actually served to create disincentives for foreign-student applicants. Second, the federal government needed to work in concert with academic institutions and large business organizations to develop a more comprehensive and systematic student recruitment program, complete with financial aid packages, better conduits for the dissemination of information, and a strong welcoming message for qualified applicants. Finally, it was argued, all interested parties should coordinate resources and share ideas in developing a more effective strategic plan overall for enhancing American competitiveness in the international student arena.

An often-cited example of self-defeating policy was the rule that in order to qualify for consideration, the aspiring student had to demonstrate intent to return eventually to his or her home country. Thus, if a student seeking a visa imprudently disclosed that he or she hoped to remain and seek employment in the U.S. after completing a degree program, that admission alone could suffice as grounds for a consular official to deny the applicant a visa. Yet, it might well be in the best interest of the United States to encourage a talented degree-holder to *remain* in the host country and thereby add to the nation's store of valuable human capital. The last thing the U.S. needed to do, ran critics' arguments, was to deny talented persons an opportunity to take up permanent residence if they wished to do so.

Other examples of immigration policy confusion were not difficult to identify. A rule was proposed by the State Department, for example, that disallowed student entry to the United States for the purpose of completing an internship. Yet, it was revealed weeks later that a high-ranking official in the same agency who had gone on record calling international exchange an "effective public-diplomacy tool" had never even heard of the internship rule.

Again, critics could point to the example where the Social Security Administration enacted measures to restrict foreigners' access to social security numbers. It was pointed out that a foreign national lacking a Social Security card would find it immeasurably more difficult to function in American society—to lease an apartment, open a bank account and so on—without the level of identification a social security card affords. At the same time, the State Department was attempting to send out its message that the United States was a friendly, welcoming place for foreign students. Marlene Johnson commented, "We cannot be successful in the intense competition for international talent if the right hand does not know what the left hand is doing, if articulated policy is cancelled out by bureaucratic actions, and if we do not bring our laws into conformity with our interests."[22]

Needed reforms most often mentioned included removal of the limit on visas granted to skilled foreign workers; a general simplification and streamlining of the application process so that a student could obtain a visa in time to meet an enrollment deadline at a college or university, special dispensation for distinguished scholars invited to the U.S. for academic purposes, and rule changes that would allow foreign students in good standing to leave the country when classes were not in session with the assurance they would be permitted to return when classes resumed.

Also cited were INS regulations governing foreign dependents. A married student, it was said, should be allowed to have his or her family members accompany him or her for as long as the primary visa-holder was enrolled in school and registering academic progress. Dependents should be allowed to seek employment and stringent time limits on their U.S. residency made more flexible as circumstances might warrant.

Yet another example of confusion in immigration policy was illustrated by a rule that forbade granting a visa to anyone whose sole purpose in seeking entry to the U.S. was to study the English language. The outcome was that many intensive English programs, existing to help students attain the language proficiency needed to enter regular degree programs, were forced to close their doors. Commenting on this, the NAFSA June 2006 position paper's authors observed, "One would be hard pressed to think of another major power in the world that discourages the study of its language."[23]

The U.S. visa system, Johnson concluded, "should be as much a gateway for international talent as it is a barrier to international criminals."[24] America, it was said, had "overcorrected" its visa system in response to the threat of terrorism; and a series of new restrictions had made it exponentially more difficult to gain entry to the United States. Unintentionally, as a result, the message sent out to prospective students overseas—all disclaimers to the contrary notwithstanding—was that they were not really wanted.

Global Competition

By the mid-2000s, if not well before, the realization began to dawn that America was not the only country in the world seeking to attract foreign students to its shores. On the global education market, the United States was still taking in the largest absolute number of foreign students. But, as noted earlier, its market share of the total number of students was beginning to slip.[25] Well before the turn of the new century, America's traditional competitor nations had begun implementing strategies for capturing a larger share of the market—necessarily at the expense of the United States.

The United Kingdom, for example, had increased its enrollments by 118,000 since 1999 and had announced in 2006 that it intended to expand its enrollments by another 100,000 within the next five years. The key to its success has not been a mystery. Today students from the European Economic Area (EEA) and Switzerland are allowed to live, study, work, and gain permanent residence anywhere in the United Kingdom. All foreign students are furnished with a permit for the entire duration of their studies plus another four months, during which time their holders may seek employment without a work permit. Students with dependents are allowed to bring their families with them and they are allowed to seek employment if they wish to work.

The European Higher Education Area (EHEA), encompassing the European Union and other states, with 45 members in all, already offers a near-seamless higher education system, one with powerful incentives built in for students to remain within the EHEA. Inducements include subsidized tuition, need-based vouchers, low-cost loans, generous scholarships in some fields, and tuition exchanges. A lower cost of living also persuades many European students to remain close to home rather than applying for admission to the more expensive programs offered in the United States.

Meanwhile, New Zealand, Japan, Australia, and Canada, among others, have begun aggressive recruitment efforts of their own, initiatives that have already translated into increased market share. Smaller states such as Qatar, Singapore, and Malaysia have followed suit, reportedly mounting successful recruitment programs of their own or in cooperation with academic institutions in developed countries. As for China and India, long-time major contributors to the U.S. enrollment of overseas students, they have begun to develop programs that carry significant rewards for students who opt to pursue their studies domestically instead of going abroad.

The Need for Strategic Planning

Internationalist agencies such as NAFSA and the IIE are agreed that international competition for students is likely to intensify in the near future. If the United States is to keep pace on a global scale, it will need to borrow and adapt from the best features of the recruitment efforts already mounted by foreign competitors: increased financial support to offset tuition fees and a relatively high cost of living, post-graduation employment opportunities, flexible loans, restructured visa procedures, reforms in immigration law, curricular and instructional innovation, and effective marketing. Above all, the paramount need will be a strategic plan that touts a welcoming environment while celebrating the advantages and prestige of an American education.

Achieving such an ambitious recruitment program would undoubtedly require close cooperation by all of the disparate governmental offices, bureaus, desks, and programs that share responsibility for the recruitment, admission, monitoring, and servicing of foreign students. "Given the internationalization of higher education and increasing competition for foreign talent," Jeanne Batalova observed," the United States will need to play to its strengths and be flexible enough to adjust its course of action when needed."[26] NAFSA, for its part, issued a clarion call to action in its *Restoring U.S. Competitiveness* paper: "To get back on track, America needs to do better. We renew our call for national leadership to elevate international educational exchange as a national priority and to establish a national strategy to ensure that the United States can attract the best in talent from around the globe."[27]

International Students in America

Less-often mentioned in discussions about international exchange has been the question of what happens to students once they arrive on American shores. What do colleges and universities need to do to ensure that a foreign student's sojourn is a productive, pleasant experience? What hazards and obstacles exist that need to be avoided if at all possible? It may be appropriate under certain conditions to view foreign students as sources of human capital, as intellectual and social resources. Yet, it must be borne in mind that foreign students are also individual human beings, with unique backgrounds, experiences, and aspirations. Reasons why the United States might exert itself to attract them to its academic institutions are not difficult to discern. But the correlative responsibility is to maximize prospects for their success during their sojourn in America.

Recent research suggests that faculty in most schools today are properly cognizant of the advantages of having foreign students in attendance—even when in certain fields faculty may resent their dependence on large foreign student enrollments to fill their classes and populate their degree programs.[28] Faculty members as a rule appreciate opportunities to forge international ties and working relationships with peers in other countries, based initially on their association with students studying in America. They also understand how valuable it is to expose American students to persons different from them and to learn about other cultures and countries. Evidence suggests that when the college's or university's leadership makes a concerted effort to integrate foreign visitors within the school's routines and activities, and encourages the student body to do likewise, American students respond with interest and expressions of hospitality.[29]

Faculty typically attempt to render assistance to foreign students when called on to do so. But reportedly many are unaware of the special resources, offices, facilities, and programs that may exist to service the unique needs of international students on their campus. Short language courses, a campus writing center, and a counseling clinic are examples of resources available to international students at many schools.

It has been suggested that the anonymity and impersonality characteristic of a large academic institution is more likely to provoke social estrangement and alienation among foreign students than will the more intimate confines of a smaller institution. Far more important, regardless of the size of school attended, is a student's willingness to reach out to others, to initiate contacts with American peers, and to work hard in adjusting to American society.[30] The imperative for the foreign student, accordingly, is three-fold: first, to seek continuously to improve English-language proficiency if needed; second, to avoid patterns of association that might isolate the person from his or her American counterparts; and, third, to make a good-faith effort to better understand and adjust to the local culture.

Host institutions can facilitate the integration of foreign students into campus life in many specific ways: by sponsoring cultural events with a global focus, encouraging friendship clubs, buddy pairings of an American and a foreign student, host family programs, demonstrations of cooking ethnic foods, and intercultural fairs. The possibilities are almost limitless.[31]

Least often mentioned is the discrimination some foreign students sometimes encounter during their stay—most commonly by off-campus outsiders—but sometimes from students on campus, too. Some evidence suggests that Middle Eastern and African students are more likely to encounter prejudice than are those from other regions such as Southeast Asia, Oceania, Europe, South Asia, the Americas, or Europe.[32] Many factors apparently come into play in determining the types of students most likely to be discriminated against. As a rule, those whose appearance most closely resembles that of Americans (Europeans, for example) are least likely to encounter prejudicial rhetoric or behavior.

Ambiguity in the Multiple Rationales for International Exchange

No historical account of international students in U.S. colleges and universities can claim adequacy if it does not acknowledge that American motives vis-à-vis bringing foreign students to the nation's institutions of higher learning have always been mixed. It remains true to claim that it has long been considered beneficial for American students to have exposure to other peoples and cultures, as provided for by the presence of foreign collegians on campus. On the other hand, even a cursory examination of U.S. immigration law and policies indicates that encouraging foreign students to come to America has rarely been given anywhere near the highest priority. Within the public sector, state legislatures have invariably been loath to offer tuition breaks at public colleges or universities for out-of-state students, much less international ones. First priority, perhaps understandably, has practically always gone to state residents.

It is often observed that attracting international scholars and students to the United States is important because it is a proven means of "making friends." As noted before, the argument is that students return home with a better and more intimate appreciation for the United States and its people than they would otherwise have acquired. Also, so it is claimed, a network of personal connections to the nation often develops, connections that may assume special importance years later in person-to-person relations between U.S. and foreign leaders.

Less well acknowledged is the fact that less elevated political considerations sometimes assume primacy. Referring to historical developments of the postwar era, two recent commentators argue that proposals to increase international exchange were not solely altruistic. "Political, economic, and cultural motives, brought on by the Cold War, also fueled the growth of international student programs on American campuses to secure the allegiance of nations in the Middle East, Asia, Latin America, and Africa," they suggest. "American political and educational leaders hoped that on returning home, the graduates of these academic programs would act as ambassadors and spread American political and cultural mores in their respective nations—to cultivate sympathy and understanding for American values worldwide."[33] The shift from earlier years, in other words, was away from mutual understanding and goodwill and toward the exporting of American cultural values, a celebration of the successes of American capitalism, and a means for countering the Cold War propaganda of Soviet Communism.

A third rationale for policies that encourage international students to come to America is the monetary contribution they and their dependents make to the U.S. economy. Universities and local college communities benefit greatly from an influx of people whose collective expenditures nationwide run into the billions of dollars. It goes without saying that monetary considerations ought not to drive U.S. policy exclusively. Still, if it can be shown that the authentic benefits of international exchange are cultural, social and political as much as they are economic, the case for promoting exchanges appears to be all the more compelling.

Recent years have witnessed much talk about how international exchange can assist the U.S. in growing its knowledge economy. NAFSA's *Restoring U.S. Competitiveness* makes the point succinctly:

> In an era of competition for scarce global talent, the countries that draw the world's best and brightest to their universities are the countries that will have the best talent pool from which to fill their cutting-edge jobs. The countries that create the most attractive environment for the world's finest scientists will do the most to enhance their scientific leadership. Indeed, the very diversity that we gain through openness to international talent itself fuels innovation and creativity."[34]

This particular rationale is hardly unassailable, given its assumption that collegiate admission policies directly determine the size and character of a scientific and technological talent pool, or that special inducements for scientists will in fact enhance their leadership while stimulating cutting-edge technological developments in different fields. It is not clear either what might persuade the so-called "best and brightest" to immigrate to the United States from abroad, either as students or later on as established adults. Nor does the argument assign much significance to *cultural* capital—the potential contributions of composers, philosophers, poets, musicians, artists, and so on, assuming these are regarded as important elements within the nation's knowledge economy.

Looking Ahead

The American Idea, it has been said, has inspired (and sometimes infuriated) peoples around the world. In an era of growing globalization and increasing international interdependence, powerful forces are being unleashed that promise to bring about sweeping, unprecedented changes in the world—social, cultural, economic, ecological, religious, and political. And in this century it may well prove to be the case that some other idea, not the American one, will occupy the world's center stage.

This much is unequivocally true: comfortable isolationism is no longer a viable option for any people or nation on the face of the planet. Technological innovation has deprived everyone of the luxury of being left alone—although through inaction on our part we may well find ourselves left behind. The inescapable truth of the matter is that all of us, around the globe, are now more and more engaged with one another. The only alternative before us is to embrace large-scale change with as much wisdom and careful thought as we can muster.

Modern modes of communication and transportation practically guarantee the development of a more tightly intertwined world economy in which we will work and live for the remainder of our lives. Meanwhile, ancient ideas and traditional values have begun to slip away under the pressure of advancing new ones. In a world of great danger and even greater promise, what is needed as much as anything else is an education that will help us achieve a measure of understanding of the changes now sweeping the planet, even while it guides us toward an uncertain future.

"We cannot be successful in the intense competition for international talent if the right hand does not know what the left hand is doing, if articulated policy is cancelled out by bureaucratic actions, and if we do not bring our laws into conformity with our interests."
Marlene M. Johnson, "Toward a New Foreign-Student Strategy,"
the *Chronicle Review* 52 (July 25, 2006): B16

"Given the internationalization of higher education and increasing competition for foreign talent, the United States will need to play to its strengths and be flexible enough to adjust its course of action when needed."
Jeanne Batalova, "The 'Brain Gain' Race Begins with Foreign Students,"
Migration Policy Institute (January 1, 2007)

Notes

Introduction

1. Cora Du Bois, *Foreign Students and Higher Education in the United States* (Washington, DC: American Council on Education, 1956), p. 1.
2. Oliver J. Caldwell, "Education Comes of Age Around the World," in *Governmental Policy and International Education*, ed. Stewart Frazer (New York: John Wiley & Sons, 1965), p. 69.
3. Robert E. Speer, "Foreword," in *The Foreign Student in America*, ed. W. Reginald Wheeler, Henry H. King, and Alexander B. Davidson (New York: Association Press, 1925), p. v.
4. Helen I. Clarke and Martha Ozawa, *The Foreign Student in the United States* (Madison, WI: School of Social Work, University of Wisconsin, 1970), p. 36.
5. Ina Coriane Brown, "The Cultural Background of International Education," in Frazer, *Governmental Policy*, p. 48.
6. Ibid., p. 52.
7. Helen C. White, "An Integrated Cultural Relations Program," in *Special Report, Conference on International Student Exchanges, May 10, 11, 12, 1948, University of Michigan*, ed. Laurence Duggan (New York: Institute of International Education, September 22, 1948), p. 4.
8. C. Selltiz and S. W. Cook, "Factors Influencing Attitudes of Foreign Students Toward the Host Country," *Journal of Social Issues* 18 (1962): 17–23, quoted in Otto Klineberg, "Psychological Aspects of Student Exchange," in *Students as Links Between Cultures*, ed. Ingrid Eide (Oslo: UNESCO and the International Peace Research Institute, 1970), p. 50.
9. Quoted in Mayr Yv DeWitt, "This Is What They Say," *The Unofficial Ambassadors 1952* (New York: Committee on Friendly Relations Among Foreign Students, 1952), p. 9.
10. Quoted in Wheeler et al., *The Foreign Student*, p. 140.
11. Clarke and Ozawa, *The Foreign Student*, pp. 33–34.
12. Thomas Marshall, "The Strategy of International Exchange," in Eide, p. 15.
13. DeWitt, "This Is What They Say," p. 11.
14. Ibid.
15. Klineberg, "Psychological Aspects," p. 43.
16. Brown, "The Cultural Background," p. 42.
17. *Global Opinion, The Spread of Anti-Americanism* (Washington, DC: Pew Research Center, January 24, 2005).

18. Pew Global Attitudes Project, *U.S. Image Up Slightly but Still Negative* (Washington, DC: Pew Research Center, June 23, 2005).
19. Geoffrey Cowan, "Hughes Offers Steps, Not Spin," *USA Today*, September 29, 2005), 12A. See also Eric Kronenwetter, "U.S. Remains Unpopular Despite Efforts to Repair Image Abroad," *International Educator* 14, no. 4 (July–August, 2005): 8.
20. Institute of International Education, *Education for One World, Annual Census of Foreign Students in the United States, 1951–52* (New York: Institute of International Education, 1952), pp. 46–7; and Institute of International Education, *Open Doors 2004, International Students in the U.S.* (New York: Institute of International Education, 2004).
21. Edward Charmwood Cieslak, *The Foreign Student in American Colleges* (Detroit, MI: Wayne University Press, 1955), p. 10.
22. Cieslak, *The Foreign Student*, p. 15.
23. Howard E. Wilson, "Foreword" to Du Bois, *Foreign Students and Higher Education*, pp. vi–vii.
24. Joe W. Neal, "The Office of the Foreign Student Adviser," *Institute of International Education News Bulletin* 17, no.5 (February 1, 1952): 37.
25. Paul M. Chalmers, "The Founding and Growth of NAFSA," *Institute of International Education News Bulletin* 26, no. 7 (April 1, 1951): 32.
26. Institute of International Education, *Education for One World*, pp. 46–47.
27. NAFSA: Association of International Educators, *International Student Handbook* (New York: Association of International Educators), chap. 1.
28. Philip G. Altbach, "Higher Education Crosses Borders," *Change* 36, no. 2 (March–April 2004): 18, 19.
29. Marlene Johnson, "Transforming the National Consciousness," *International Educator* 14, no. 5 (September–October, 2005): 37.

Chapter 1

1. Quoted from Plato's *Protagoras* in John W. H. Walden, *The Universities of Ancient Greece* (New York: Charles Scribner's Sons, 1909), pp. 16–17.
2. Hellenic Ministry of Culture, University of Patras, Athens, 1995–2001. http://www.culture.gr (accessed January 1, 2007).
3. W. W. Capes, *University Life in Ancient Athens* (New York: G. E. Stechert and Company, 1922), p. 28.
4. Ibid., p. 27.
5. H. I. Marrou, *A History of Education in Antiquity* (New York: Mentor, 1964), p. 399.
6. Lloyd W. Daly, "Roman Study Abroad," *American Journal of Philology* 71 (1950): 53. Daly uses Strabo as a source.
7. Marrou, *History of Education*, p. 260.
8. Preston Chesser, *The Burning of the Library of Alexandria*, http://www.ehistory.com/world/articles (accessed January 1, 2007).
9. Marrou, *History of Education*, p. 261.

10. Ibid., p. 294.
11. Walden, *Universities*, pp. 313–14.
12. Charles Homer Haskins, *The Renaissance of the Twelfth Century* (Cambridge, MA: Harvard University Press, 1928), p. 372.
13. Ibid., p. 394.
14. Ibid., pp. 395–96.
15. Helen Waddell, *The Wandering Scholars* (London, 1927), p. 164.
16. J. A. Symonds, *Wine, Women and Song* (London: private reprint, 1884).
17. Haskins, *The Renaissance*, p. 82.
18. Ibid., pp. 82–83.
19. John Herman Randall, *The Making of the Modern Mind* (Cambridge, MA.: Riverside Press, 1940), p. 116.
20. Christopher J. Lucas, *American Higher Education: A History*, 2nd ed. (New York: Palgrave Macmillan, 2006), pp. 41–42.
21. Haskins, *The Rise of Universities* (New York: Henry Holt and Company, 1923), p. 13.
22. John W. Baldwin, *The Scholastic Culture of the Middle Ages, 1000–1300* (Long Grove, IL: Waveland Press, 1997), p. 46.
23. Gabriel Compayre, *Abelard and the Origin and Early History of Universities* (New York: Charles Scribner's Sons, 1910), pp. 272–77; Pearly Kibre, "Nations in the Mediaeval Universities," *Mediaeval Academy of America* 49 (1948): pp. 101–02.
24. Lucas, *American Higher Education*, p. 44.
25. Haskins, *The Rise of Universities*, p. 8.
26. Marjorie Rowling, *Life in Medieval Times* (New York: Berkley Publishing Group, 1979), p. 148.
27. Lynn Thorndike, *University Records and Life in the Middle Ages* (New York: Columbia University Press, 1944), pp. 260–61, 332, 352–53.
28. Lucas, *American Higher Education*, p. 49.
29. Baldwin, *The Scholastic Culture*, p. 50.
30. Charles T. Wood, *The Quest for Eternity, Medieval Manners and Morals* (New York: Doubleday Anchor, 1971), pp. 156–57.
31. Lucas, *American Higher Education*, p. 50.
32. Ibid., p. 84.
33. Ibid., p. 93.
34. Quoted in Felix Markham, *Oxford* (London: Werden Feld and Nicolson, 1967), p. 84.
35. Sara Douglass, *Pilgrims, Students, and Gentlemen* (2000). http://www.saradouglass.com/1570.html (accessed January 1, 2007).
36. Louis B. Wright, *Middle Class Culture in Elizabethan England* (1935; repr. London, 1964), pp. 44–48.
37. Richard S. Lambert, *Grand Tour: A Journey in the Tracks of Aristocracy* (New York: E. P. Dutton & Company, 1937); and Jeremy Black, *The British and the Grand Tour* (Worcester, UK: Billing and Sons, 1985).
38. Lambert, *Grand Tour*, p. 18. Consult also Joan Simon, *Education and Society in Tudor England* (Cambridge: Cambridge University Press, 1966).

39. Lambert, *Grand Tour*, p. 18.
40. Ibid., p. 18.
41. Ibid., p. 19.
42. Christopher Hibbert, *The Grand Tour* (New York: G. P. Putnam's Sons, 1969), p. 10.
43. William E. Mead, *The Grand Tour in the Eighteenth Century* (New York: Benjamin Bloom, 1972), p. 231.

Chapter 2

1. William W. Brickman, "Historical Development of Governmental Interest in International Higher Education," in *Governmental Policy and International Education*, ed. Stewart Fraser (New York: John Wiley & Sons, Inc. 1965), p. 29.
2. W. Reginald Wheeler, Henry King, and Alexander Davidson, *The Foreign Student in America* (New York: Association Press, 1925), p. 7.
3. Ibid., p. 11.
4. Thomas Jefferson, *Writings: Autobiography, Notes on the State of Virginia, Public and Private Papers, Addresses, Letters*, ed. Merrill D. Peterson (New York: Library of America, 1994).
5. Brickman, "Historical Development," p. 30.
6. George Washington, The writings of George Washington from the original manuscript sources, 1732–1799. Electronic Text Center, University of Virginia Library. http://etext.virginia.edu/toc/modeng/public/WasFi34.html (accessed January 1, 2007).
7. Brickman, "Historical Development," p. 30.
8. Stewart Fraser and William W. Brickman, *A History of International and Comparative Education* (Chicago: Scott, Foresman and Company, 1968), p. 294.
9. Richard Hofstadter and Wilson Smith, eds., *American Higher Education: A Documentary History* (Chicago: University of Chicago Press, 1961); also see Frederick Rudolph, *The American College and University* (Athens: University of Georgia Press, 1990), p. 42.
10. Wheeler et al., *The Foreign Student*, p. 7.
11. Warner Berthoff, "George Ticknor: Brief Life of a Scholarly Pioneer 1791–1871," *Harvard Magazine* (January–February 2005).
12. Ibid.
13. Fraser and Brickman, *A History*, pp. 293–295.
14. Ibid.
15. Ibid.
16. Isaac Fidler, *Observations on Professions, Literature, Manners and Emigration in the United States and Canada Made During a Residence There in 1832* (London: Whittaker, Treacher, and Company, 1833), p. 46.
17. Ibid. p. 47.
18. Fraser and Brickman, *A History*, p. 367.

19. Ibid., p. 424.
20. Ibid., p. 426.
21. Michael E. Sadler, "Impressions of American Education," *Educational Review* 25 (March 1903): 217–231.
22. Ibid.
23. James Biggs, *History of Don Francisco de Miranda's Attempt to Effect a Revolution in South America*; biography by W. S. Robertson (Lodon, Printed for the author by T. Gillet, 1809), xv, 312, p. 23.
24. Ibid.
25. University of Virginia, *UVA Names Its Spanish House in Honor of Alumnus Fernando Bolivar, Nephew and Adopted Son of South American Hero Simon Bolivar*. http://www.virginia.edu/topnews/textonlyarchive/November_1996/bolivar.txt (accessed January 1, 2007).
26. Ibid.
27. Columbia University, "Mario Garcia Menocal," *The Columbia Encyclopedia*, 6th ed. (New York: Columbia University Press, 2001).
28. Qian Ning, *Chinese Students Encountering America*, trans. T. K. Chu (Seattle: University of Washington Press, 2002).
29. Harold Hongju Koh, "Yellow in a White World," *Yale Global*, September 27, 2004. Yale Center for the Study of Globalization. http://yaleglobal.yale.edu/display.article?id=4653 (accessed January 1, 2007).
30. Ning, *Chinese Students*, p. 12.
31. Yung Wing, *My Life in China and America* (New York: Henry Holt & Co., 1909), p. 14.
32. Ibid., p. 20.
33. E. J. M Rhodes, "In the Shadow of Yung Wing," *Pacific Historical Review* 74 (February 2005): 19–58.
34. Ning, *Chinese Students*, p. x.
35. Ibid.
36. Yung Wing, *My Life in China*, p. 69.
37. Ibid., p. 71.
38. Ning, *Chinese Students*, p. x.
39. Judith Ann Schiff, "When East Meets West," *Yale Alumni Magazine* (November/ December 2004), Yale University Library.
40. Steve Courtney, "Rising Power in the East: Joe Twichell, Mark Twain and the Chinese Educational Mission, 1872–81." Unpublished research (Hartford, CT: Asylum Hill Congregational Church archives. http://www.ahcc.org/TWAIN/MarkJoeandBOYS4.htm (accessed January 1, 2007).
41. Ibid.
42. Ibid.; also see William Hung, "Huang Tsun-Hsing's Poem 'The Closure of the Education Mission in America,'" *Harvard Journal of Asiatic Studies* 18, nos. 1–2 (June 1955): 50–73.
43. Timothy Kao, *Yung Wing (1828–1912) and Young Chinese Students in America (1872–1881)*, http://www.120chinesestudents.org/yung.html (accessed January 1, 2007).
44. Hung, "Huang Tsun-Hsien's Poem."

45. Ibid.
46. Courtney, "Rising Power in the East."
47. Ibid.
48. Edward Charnwood Cieslak, *The Foreign Student in American Colleges* (Detroit, MI: Wayne University Press, 1955).
49. Ning, *Chinese Students*, p. xi.
50. Teng Ssu-you and J. K. Fairbank, *China's Response to the West: A Documentary Survey, 1839–1923* (Cambridge, MA: Harvard University Press, 1954).
51. Hui Huang, "Overseas Studies and the Rise of Foreign Cultural Capital in Modern China," *International Sociology* 17, no. 1 (2002): 35–55.
52. Kao, *Yung Wing*.
53. Yung Wing, *My Life in China*, p. 204.
54. Ray Moore, *Amherst College and Japan*, Amherst College of Asian Languages and Civilizations, 2004. http://www.amherst.edu/~asian/yushien.html (accessed January 1, 2007).
55. Jerome Dean Davis, *A Sketch of the Life of Reverend Joseph Hardy Neesima* (Kyoto, Japan: Doshiva University, 1936).
56. Janet E. Goff, "Tribute to a Teacher: Uchimura Kanzo's Letter to William Smith Clark," *Monumenta Nipponica* 43, no. 1 (Spring 1988): 95–100.
57. Moore, *Amherst College*.
58. Marion Smith, "Overview of INS History," in *A Historical Guide to the U.S. Government*, ed. George T. Kurian (New York: Oxford University Press, 1998).
59. Ibid.
60. Ibid.
61. Peggy Pascoe, "At America's Gates: Chinese Immigration during the Exclusion Era, 1882–1943 (review)," *Journal of Social History* 38, no. 3 (Spring 2005): 812–814.
62. Smith, "Overview of INS History."
63. Ibid.
64. Thomas M. Pitkin, *Keepers of the Gate: A History of Ellis Island* (New York: University Press, 1975).

Chapter 3

1. Frederick Rudolph, *The American College and University: A History* (Athens and London: University of Georgia Press, 1990), p. 357.
2. Edward Charnwood Cieslak, *The Foreign Student in American Colleges* (Detroit, MI: Wayne University Press, 1925), p. 9.
3. W. Reginald Wheeler, Henry H. King, and Alexander B. Davidson, *The Foreign Student in America* (New York: Association Press, 1925), p. 9.
4. Ibid., p. 11.
5. Ibid., p. 12.

6. George Zook, "The Residence of Foreign Students at American Universities." *United States Bureau of Education, Bulletin No. 27* (1915), whole no. 654.
7. Wheeler et al., *The Foreign Student*, p. 13.
8. Ibid., p. 14.
9. Stacey Bieler, *Patriots or "Traitors"?: A History of American Educated Chinese Students* (Armonk, NY: M. E. Sharpe, 2003); also see Richard Hooker, *Ch'ing China: The Boxer Rebellion*, Washington State University. http://www.wsu.edu/~dee/CHING/BOXER.htm (accessed January 1, 2007).
10. Ibid.
11. Isaac Kandel, United States Activities in International Cultural Relations, *American Council on Education Studies*, series I, IX, no. 23: 2.
12. Wheeler et al., *The Foreign Student*, p. 9.
13. John Fryer, "Admission of Chinese Students to American Colleges," *United States Bureau of Education*, Bulletin no. 2 (1909), whole no. 399. Washington, DC: Government Printing Office.
14. Ibid., pp. 199–200.
15. Ibid., p. 215.
16. Wheeler et al., *The Foreign Student,*, p. 15; also see Bieler, *Patriots or "Traitors"?*.
17. Tsinghua University, Official Web Site, Beijing. http://www.tsinghua.edu (accessed January 1, 2007).
18. Ibid.
19. Isaac L. Kandel, "United States Activities in International Cultural Relations," *Council on Education Studies*, series I, IX, no. 23: 53.
20. Institute of International Education, *Blueprint for Understanding* (New York: Author, 1949).
21. Columbia University Center on Chinese Education Web Site. http://www.TC.columbia.edu/centers/coce (accessed January 1, 2007).
22. Yi Zhuxian, "Hu Shi, Chinese and Western Culture," *The Journal of Asian Studies* 51, no. 4 (November 1992): 915–916; also see Irene Eber, *Hu Shi*, Claremont Graduate University, http://www.cgu.edu/pages/3279.asp (accessed January 1, 2007).
23. Bieler, *Patriots or "Traitors"?*, p. 200.
24. Chaing Yung-chen, *Chinese Studies in History* 36, no. 3 (Spring 2003): 38–62.
25. Ibid., p. 40.
26. Ibid., p. 41.
27. Ibid., p. 41.
28. Ibid., p. 48.
29. Smithsonian Asian Pacific American Program, *The Filipino American Story*. http://www.apa.si.edu/filamcentennial/timeline.html (accessed January 1, 2007).
30. Susan Evangelista, *Carlos Bulosan and His Poetry: A Biography and Anthology* (Seattle: University of Washington Press, 1985).
31. Stanley Karnow, *In Our Image: America's Empire in the Philippines* (New York: Random House, 1990).

32. Barbara M. Posadas, *The Filipino Americans*, "The New Americans Series," series ed. Ronald H. Bayor (Westport, CT: Greenwood Publishing Group, 1999).
33. Wheeler et al., *The Foreign Student*, p. 17.
34. Posadas, *The Filipino Americans*, p. 3.
35. Anna Liza R. Ong, *First Filipino Women Physicians*. Society of Philippine Health History, 2004. http://www.doh.gov.ph/sphh/filipino/women.htm (accessed January 1, 2007).
36. Ibid.
37. Wheeler et al., *The Foreign Student*, p. 20.
38. Eduardo Bananal, *Camilo Osias: Educator and Statesman* (Quezon City, PI: Manlapaz Publishing Co., 1974).
39. Ibid.
40. Wheeler et al., *The Foreign Student*, p. 20.
41. Ibid.
42. Anil K. Bera, "Cosmopolitan Club, Tagores, and UIUC: A Brief History of 100 Years." https://netfiles.uiuc.edu/ro/www/CosmopolitanClubattheUniversity ofIllinois.html (accessed January 1, 2007).
43. Ibid.
44. Stewart Fraser, *Governmental Policy and International Education* (New York: John Wiley & Sons, 1964), p. 92.
45. Louis B. Lochner, "Cosmopolitan Clubs in American University Life," *American Review of Reviews* 37 (1908): 1.
46. Alan K. Laing, *History of the Cosmopolitan Club*. University of Illinois, http://www.apa.si.edu/filamcentennial/timeline.html (accessed January 1, 2007).
47. L. Glazier and L. Kenschaft, "Welcome to America," *International Educator* 11, no. 3 (Summer 2002): 7.
48. Ibid., p. 8.
49. Ibid.
50. Ibid., p. 9.
51. Ibid.
52. University of California at Berkeley Web Site, "International House Berkeley Historical Background 2000." http://ias.berkeley.edu/ihouse/I/history.html (accessed January 1, 2007).
53. Committee on Friendly Relations Among Foreign Students, *Unofficial Ambassadors* (New York: Author, 1945), p. 1.
54. David Rockefeller, *Memoirs* (New York: Random House, 2002), pp. 124–25.
55. University of California at Berkeley, "International House Berkeley."
56. Committee on Friendly Relations, *Unofficial Ambassadors*.
57. Analee Allen, "International House Marks Seventieth Year," University of California Berkeley Web Site. http://ias.berkeley.edu/ihouse/I/tribune.html (accessed January 1, 2007).
58. African American Registry Web Site. http://www.aaregistry.com/african_american_history/1138/Delialah_L_Beasley (accessed January 1, 2007).
59. University of California Berkeley, "International House Berkeley."

60. Ibid.
61. Evan Ross, "Friendships Traded in Gaps of War, Ideology International House History Before, After Wartime," *The Daily Californian*, 2005. dailycalifornian @dailycal.org (obtained January 1, 2007).
62. J. Ariff, "Oral Interview with A. C. Blasdell (1928–1961)," *Foreign Students and Berkeley International House* (Berkeley: University of California Press, 1967).
63. Ibid.
64. Fraser, *Governmental Policy*, p. 95.
65. Committee on Friendly Relations, *Unofficial Ambassadors*, p. 3.
66. Ibid.
67. Kim Clarke, "The Barbour Scholars," The Michigan Difference Web Site, The University of Michigan Office of Development, Winter 2003. http://www.giving.umich.edu/leadersbest/winter2003/barbour.htm (accessed January 1, 2007).
68. Wheeler et al., *The Foreign Student*, p. 179.
69. Ibid., p. 180.
70. Ibid., p. 182.
71. Liping Bu, "The Challenge of Race Relations: American Ecumenism and Foreign Student Nationalism, 1900–1940," *Journal of American Studies* 35 (2001): 217–237.
72. Martha Lund Smalley, *Guide to the Archives of the Chinese Students' Christian Association in North America* (New Haven, CT: Yale University Library, 1983).
73. Ibid.
74. Wheeler et al., *The Foreign Student*, p. 12.
75. Ibid., p. 11.
76. Kandel, "United States Activities," p. 54.
77. *Time* Magazine, "Father and Son," November 4, 1917.
78. Cieslak, *The Foreign Student*, p. 11.
79. Ibid.
80. Fraser, *Government Policy*, p. 96.
81. Ibid., p. 97.
82. Marion Smith, "Overview of INS History," Immigration and Naturalization Services Web Site. http://uscis.gov/graphics/aboutus/history/articles/OVIEW/htm (accessed January 1, 2007).

Chapter 4

1. *Time* Magazine, "Alarums and Excursions," XXXIV, no. 12 (September 18, 1939). http://www.time.com (accessed January 1, 2007).
2. *Time* Magazine, "Long View," XLIII, no. 13 (March 27, 1944). http://www.time.com (accessed January 1, 2007).
3. Stewart Fraser, *Governmental Policy and International Education* (New York: John Wiley & Sons, 1964).
4. Bill Clinton, *My Life* (New York: Alfred A. Knopf, 2004), p. 98.

5. Ibid., p. 99.
6. Institute of International Education, *Fulbright Program for U.S. Students*. Washington, DC, 2007. http://us.fulbrightonline.org/about_programhistory.html (accessed January 1, 2007).
7. Richard T. Arndt and David Lee Rubin, eds., *The Fulbright Difference 1948–1992: Studies on Cultural Diplomacy and the Fulbright Experience* (New Brunswick, NJ: Transaction Publishers, 1993).
8. Ibid., p. 12.
9. S. J. Niefeld and Harold Mendelsohn, "How Effective Is Our Student-Exchange Program?" *Educational Research Bulletin* 33, no. 2 (February 10, 1954): 29–37.
10. Arndt and Rubin, *The Fulbright Difference*, p. 13.
11. University of Arkansas, Fulbright College of Arts and Sciences, Fulbright Exhibit. http://www.uark.edu/depts/speccoll/fulbrightexhibit/bi1apic.html (accessed January 1, 2007).
12. Arndt and Rubin, *The Fulbright Difference*, p. 14.
13. Ibid.
14. Ibid., p. 15.
15. University of Arkansas, Fulbright College of Arts and Sciences.
16. Isabel A. Maurer, "The Fulbright Act in Operation," *Far Eastern Survey* 18, no. 9 (May 4, 1949): 104–07.
17. Institute of International Education, "Timeline and Institute Highlights." http://www.iie.org (accessed January 1, 2007).
18. Pamela Spence Richards, "Cold War Librarianship: Soviet and American Library Activities in Support of National Foreign Policy, 1946–1991," *Libraries & Culture* 36, no. 1 (Winter 2001): 183–203.
19. Ibid.
20. *Time* Magazine, "The Man in the Window," LIII, no. 1 (January 3, 1949). http://www.time.com (accessed January 1, 2007).
21. Ibid.
22. Ann Coulter, "Are You Now or Have You Ever Been a Second-Rate Filmmaker?", Townhall.com (November 17, 2005). http://www.townhall.com/columnists/AnnCoulter/2005 (accessed June 1, 2007).
23. *Time* Magazine, "The Man in the Window."
24. United Nations Educational, Scientific and Cultural Organization, *The Organization's History*. http://www.UNESCO.org.
25. Ibid.
26. *Time* Magazine, "E.R.P. and M.I.T.," LVI, no. 5 (July 31, 1950). http://www.time.com (accessed January 1, 2007).
27. Institute of International Education, *Open Doors 1950, Report on International Education Exchange* (New York: Author, 1950).
28. Committee on Friendly Relations Among Foreign Students, *The Unofficial Ambassadors* (New York: Author, 1942).
29. Institute of International Education, *Open Doors 1949, Report on International Education Exchange* (New York: Author, 1949).
30. Ibid., p. 7.
31. Ibid., p. 21.

32. Ibid., p. 17.
33. Committee on Friendly Relations, *The Unofficial Ambassadors*.
34. Fraser, *Governmental Policy*, p. 99.
35. Ibid.
36. Ibid., p. 102.
37. Ibid., p. 102.
38. Ibid., p. 103.
39. Edward C. Cieslak, *The Foreign Student in American Colleges* (Detroit, MI: Wayne University Press, 1955), p. 15.
40. Fraser, pp. 105–06.
41. Cieslak, pp. 39–60.
42. Ibid.
43. Cieslak, *The Foreign Student*, pp. 38–65.
44. Ibid.
45. Jana Braziel, "History of Migration and Immigration Laws in the United States," University of Massachusetts Web Site. http://www.umass.edu/complit/aclanet/USMigrat.html (accessed January 1, 2007).
46. Committee on Educational Interchange Policy, *Chinese Students in the United States, 1948–1955* (New York: Author, 1956), p. 3.
47. Ibid., pp. 4–5.
48. Ibid.
49. United States House of Representatives, Report No. 1039, "Relief of Chinese Students," July 13, 1949, p. 2.
50. Committee on Educational Interchange Policy, *Chinese Students*, p. 7.
51. Ibid., p. 10.

Chapter 5

1. John Woolley and Gerhard Peters, *The American Presidency Project*. University of California, Santa Barbara. http://www.presidency.ucsb.edu (accessed January 1, 2007).
2. Bill Clinton, *My Life* (New York: Alfred A Knopf, 2004), p. 100.
3. Ellen Schrecker, *The Age of McCarthyism* (New York: Bedford/St. Martin, 2005).
4. Nathan J. Heller and Jessica R. Rubin-Wills, *Crimson* Staff Writers, "Trying Times, Harvard Takes Safe Road," *The Harvard Crimson*, June 5, 2003.
5. Ibid.
6. Institute of International Education, *Open Doors 1950, Report on International Education Exchange* (New York: Author, 1951), p. 9.
7. Ibid.
8. Ibid.
9. M. Brewster Smith, "Features of Foreign Student Adjustment," *Journal of Higher Education* 26, no. 5 (1955): 234.
10. *Time* Magazine, "Keeping a Pledge," LXV, no. 15 (April 11, 1955). http://www.time.com (accessed January 1, 2007).
11. Institute of International Education, *Open Doors 1950*, p. 11.

12. Ibid., p. 12.
13. Ibid.
14. Committee on Educational Interchange Policy, *Chinese Students in the United States, 1948–55* (New York: Author, 1956).
15. Ibid., p. 10.
16. *Time* Magazine, "The Confidence Game," LXVIII, no. 2 (July 9, 1956). http://www.time.com (January 1, 2007).
17. Ibid.
18. Committee on Educational Interchange Policy, *Chinese Students*, p. 11.
19. *Time* Magazine, "The Confidence Game."
20. Committee on Educational Interchange Policy, *Chinese Students*, p. 12.
21. Ibid., p. 13.
22. Committee on Educational Interchange Policy, *Expanding University Enrollments and the Foreign Student* (New York: Author, 1957), p. 2.
23. Ibid., p. 2.
24. Ibid., p. 4.
25. Ibid., p. 2.
26. Ibid.
27. Ibid.
28. *Time* Magazine, "Anti-Homusicku," LX, no. 7 (August 18, 1952). http://www.time.com (accessed January 1, 2007).
29. Ibid.
30. Institute of International Education, *Open Doors 1955, Report on International Education Exchange* (New York: Author, 1955).
31. Ibid.
32. Ibid.
33. Ibid.
34. *Time* Magazine, "A Doctor for Kenya," LXXIV, no. 3. http://www.time.com (accessed January 1, 2007).
35. Institute of International Education, *Open Doors 1955–1956*.
36. Ibid.
37. Durant da Ponte, "The Fulbright Program in Spain," *Hispania* 44, no. 3 (September 1961): 473–75.
38. Council on International Education Exchange, *CIEE History Part One: 1947–1960*. http://www.ciee.org/about/downloads/history1.pdf (accessed January 1, 2007).
39. Ibid.
40. Ibid., p. 12.
41. AMIDEAST, "1951–2006: 50 Years of Bridging Cultures and Building Understanding." Washington, DC. http://www.amideast.org/about/50_years/default.htm (accessed January 1, 2007).
42. Ibid.
43. Institute of International Education, *Open Doors 1959*.
44. Ibid.
45. Ibid.
46. Ibid.

Chapter 6

1. John Woolley and Gerhard Peters, *The American Presidency Project*, John F. Kennedy, Remarks at a Reception for Foreign Students on the White House Lawn, May 10, 1962. University of California, Santa Barbara. http://www.presidency.ucsb.edu (accessed January 1, 2007).
2. Ibid.
3. Institute of International Education, *Open Doors 1961, Report on International Education Exchange* (New York: Author, 1961), p. 7.
4. Ibid., p. 8.
5. The John F. Kennedy Presidential Library and Museum, Image Archives. August 8, 1961, Memorandum to the Secretary of State. http://www.jfklibrary.org (accessed January 1, 2007).
6. Institute of International Education, *Open Doors 1961*.
7. John Woolley and Gerhard Peters, *The American Presidency Project*, Lyndon B. Johnson, Remarks to a Group of Foreign Students, May 5, 1964, University of California, Santa Barbara. http://www.presidency.ucsb.edu (accessed January 1, 2007).
8. Woolley and Peters, *The American Presidency Project*, Lyndon B. Johnson, Special Message to the Congress Proposing International Education and Health Programs, February 2, 1966. http://www.presidency.ucsb.edu (accessed January 1, 2007).
9. *Time* Magazine, "Welcome Stranger," LXXVIII, no. 1 (September 8, 1961). http://www.time.com/time/magazine/article/0,9171,872748,00.html (accessed January 1, 2007).
10. Ibid.
11. Ibid.
12. George M. Houser, "Meeting Africa's Challenge: The Story of the American Committee on Africa," *Issue: A Journal of Opinion* 6, nos. 2–3 (Summer–Autumn 1976): 16–26.
13. *Time* Magazine, "The African Question," LXXVI, no. 9 (August 29, 1960). http://www.time.com (accessed January 1, 2007).
14. Ibid.
15. Ibid.
16. Julien Engel, "The African-American Institute," *African Studies Bulletin* 6, no. 3 (October 1963): 13–18.
17. Institute of International Education, *Opening Minds to the World*, "Timeline and Institute Highlights." http://www.iie.org (accessed January 1, 2007).
18. Ibid.
19. Hubert H. Humphrey Fellowship Program 2001–2002, "Humphrey Fellows." http://www.exchanges.state.gov/education/hhh/hhhmag01-02.pdf (accessed January 1, 2007).
20. Institute of International Education, *Opening Minds to the World*, "South Africa Education Program (SAEP)." http://www.iie.org/Content/NavigationMenu/Programs_ Portal/Archives1 (accessed January 1, 2007).

21. Ibid.
22. Ibid.
23. Ibid.
24. Institute of International Education, *Open Doors 1969, Report on International Education Exchange* (New York: Author, 1969), p. 4.
25. Ibid., p. 7.
26. Ibid., p. 9.
27. Ibid., p. 8.
28. Institute of International Education, *Open Doors 1971, Report on International Education Exchange* (New York: Author, 1971), p. 8.
29. Leonard V. Koos, "Recent Growth of the Junior College" [1928]. Reprinted in *American Journal of Education* 91, no. 4, *The First 90 Years* (August 1983): 468–79.
30. Frederick Rudolph, *The American College and University* (Athens and London: University of Georgia Press, 1990), p. 351.
31. Institute of International Education, *Open Doors 1971*, p. 8.
32. Ibid., p. 9.
33. College Entrance Examination Board, "The Foreign Student in United States Community and Junior Colleges"; A Colloquium Held at Wingspread, Racine, Wisconsin, October 18–20, 1977. ERIC: ED154858.
34. Robert L. Breuder, *A Statewide Study: Identified Problems of International Students Enrolled in Public Community/Junior Colleges in Florida*. Florida State University and the University of Florida: Center for State and Regional Leadership, May 1972/Eric Microfiche Collection No. ED 062 977.
35. College Entrance Examination Board, "The Foreign Student," p. 30.
36. Billy W. Baker, "The Foreign Transfer Student: Feel and Fact," *College and University* 50 (Spring 1975): 243–46.
37. Maxwell C. King and Robert L. Breuder, "Editor's Notes," *New Directions for Community Colleges* 1979, no. 26: vii-x. Also see CCID Web Site, "History of CCID," 1976–1992, http://www.ccid.cc/history/hist4.htm (accessed January 1, 2007).
38. S. V. Martorana, "Constraints and Issues in Planning and Implementing Programs for Foreign Students in Community and Junior Colleges," in *The Foreign Student in United States Community and Junior Colleges* (New York: College Entrance Examination Board, 1978), p. 37.
39. Ibid., p. 39.
40. Teresa Brawner Bevis, *Attitudes and Characteristics of Foreign Freshmen at the University of Arkansas*. Unpublished PhD diss., 2001, pp. 62–78.
41. Ibid.
42. Ibid.
43. Institute of International Education, *Open Doors 1979, Report on International Education Exchange* (New York: Author, 1979).
44. Robert Sedgwick, "Middle Eastern Students Find Options at Home and Elsewhere," *World Education News and Reviews* (November/December 2004).

Chapter 7

1. Geoffrey Maslan, "Big Growth in Foreign Students Predicted for Australia," *Chronicle of Higher Education* (December 15, 1995).
2. Paul Desruisseaux, "Intense Competition for Foreign Students Sparks Concerns about U.S. Standing," *Chronicle of Higher Education* 45, no. 7 (October 9, 1998): A55.
3. Ibid., p. A55.
4. U.S. Leadership in International Education, *The Lost Edge*. Conference Report and Action Agenda, United States Information Agency and Educational Testing Service. Washington, DC, 1998.
5. Paul Desruisseaux, "2-Year Colleges at Crest of Wave in U.S. Enrollment by Foreign Students," *Chronicle of Higher Education* 45, no. 16 (December 11, 1998): A66.
6. Hey-Kyung Koh, *The Impact of Community Colleges on International Education*. http://opendoors.iienetwork.org (accessed January 1, 2007).
7. Colin Woodward, "At Community Colleges, Foreign Students Discover Affordable Degree Programs," *Chronicle of Higher Education* 47, no. 12 (November 17, 2000): A77.
8. Paula Zeszotarski, Issues in Global Education Initiatives in the Community College, *Community College Review* 29 (Winter 2001).
9. American Council on International Intercultural Education and the Stanley Foundation, "Building the Global Community: The Next Step." November 28–30, 1994. http://www.stanleyfoundation.org/reports/CC1.pdf (accessed January 1, 2007).
10. Ibid.
11. Audree M. Chase and James R. Mahoney, eds., *Global Awareness in Community Colleges, A Report on a National Survey* (Washington, DC: American Association of Community Colleges, 1996), pp. v–viii.
12. George R. Boggs, "About AACC," *American Association of Community Colleges*, 2006. http://www.aacc.nche.edu.
13. American Council on International Intercultural Education, "Building the Global Community."
14. Desruisseaux, "2-Year Colleges," p. A66.
15. Ibid.
16. Daniel Golden, *Foreign Students' High Tuition Spurs Junior Colleges to Fudge Facts*, The Boston College Center for International Higher Education, Fall 2002. http://www.bc.edu/bc_org/avp/soe/cihe/newsletter/News29/text004.htm (accessed January 1, 2007).
17. Ibid.
18. Ibid.
19. Ibid.
20. Christine F. Meloni, *Adjustment Problems of Foreign Students in U.S. Colleges and Universities*. ERIC #: ED276296.

21. Philip Altbach, David Kelly, and Y. Lulat, *Research on Foreign Students and International Study* (New York: Praeger, 1985).
22. Philip Altbach and Jing Wang, *Foreign Students and International Study 1984–1988* (University Press of America, 1989).
23. Paul B. Marion, "Research on Foreign Students at Colleges and Universities in the United States," *New Directions for Student Services* 1986, no. 36: 65–82.
24. Ibid.
25. Ibid.
26. Gary Althen, *The Handbook of Foreign Student Advising* (Washington, DC: NAFSA, 1984).
27. Marion, "Research on Foreign Students," pp. 65–82.
28. Institute of International Education, *Open Doors 1996, Report on International Education Exchange* (New York: Author, 1996).
29. Marion, "Research on Foreign Students," pp. 65–82.
30. Jean M. Johnson and Mark Regets, *International Mobility of Scientists and Engineers to the United States: Brain Drain or Brain Circulation?* (Arlington, VA: National Science Foundation, 1998 [NSF 98-316]).
31. Ibid.
32. In a Letter to the Editor, *Chronicle of Higher Education* 45 (June 4, 1999): B3.
33. Johnson and Regets, *International Mobility of Scientists*.
34. Ibid.
35. Ibid.
36. Allen E. Goodman, "What Foreign Students Contribute," *Chronicle of Higher Education* 42, no. 23 (February 16, 1996).
37. Institute of International Education, *Open Doors 2000, Report on International Education Exchange* (New York: Author, 2000), p. 3.
38. Ibid., p. 8.
39. Ethiraian Anbarasan, "India's Loss, West's Gain — Higher Education . . . and After?" *UNESCO Courier* (September 1998).
40. Ibid.
41. Ibid.
42. Beth McMurtrie, "American Colleges Experience a Surge in Enrollments from India," *Chronicle of Higher Education* 48, no. 12 (December 10, 1999).
43. Anbarasan, "India's Loss."
44. Ibid.
45. VisaLaw.com, The Immigration Law Portal, "IIRIRA 96: A Summary of the New Immigration Bill." http://www.visalaw.com/96nov/3nov96.html (accessed January 1, 2007).
46. Ibid.
47. WebImmigration.com, "Students and Exchange Visitors." http://www.webimmigration.com/us/students.html (accessed January 1, 2007).
48. Amy Magaro Rubin, "U.S. Tests a New System to Track Foreign Students," *Chronicle of Higher Education* (September 19, 1997): A49.
49. Ibid.

50. John A. Bogdanski, "The IRS Imposes a Ridiculous Burden on Foreign Students," *Chronicle of Higher Education* (May 2, 1997).
51. Ibid.
52. Ibid.

Chapter 8

1. Brian Kladko, "Foreign Students Face Sharp Scrutiny," [Bergen County, NJ] *Record*, January 31, 2000. http://www.highbeam.com/DocPrint,aspx?DocId=1P1:24806528 (accessed January 1, 2007).
2. Beth McMurtrie, "Commission Seeks to Prevent Terrorism by Monitoring Foreign Students," *Chronicle of Higher Education* 46 (June 16, 2000): A49.
3. Quoted in McMurtrie, "Commission Seeks to Prevent Terrorism," p. A49.
4. McMurtrie, "Commission Seeks to Prevent Terrorism," p. A49.
5. Quoted in McMurtrie, "Commission Seeks to Prevent Terrorism," p. A49.
6. James M. O'Neill, "Foreign Students May Bear Brunt of Terrorism with Stricter Visas," Knight Ridder Tribune News Service, September 27, 2001. http://www.highbeam.com/DocPrint,aspx?Doc1G1:78713987 (accessed January 1, 2007).
7. Victor C. Johnson, "Taking Stock, Making Strides," *International Educator* 14, no. 1 (March–April, 2005): 4; Olga Bain and William K. Cummings, "Where Have the International Students Gone?" *International Instructor* 14 (March–April 2005): 20.
8. Johnson, "Taking Stock," p. 4.
9. Christopher Murphy, "Going with the Flows?" *International Educator* 14 (March–April 2005): 2.
10. Quoted in O'Neill, "Foreign Students May Bear Brunt," p. 1.
11. Ibid.
12. Ibid.
13. Sara Hebel, "Proposed Rules on Foreign Students Leave Many Colleges Worried," *Chronicle of Higher Education* (October 26, 2001): A29.
14. Cited in O'Neill, "Foreign Students May Bear Brunt," p. 1.
15. Terry Hartle, quoted in Ron Southwick, "College Officials Fear Tightening of Student-Visa Rules in Wake of Terrorist Attacks," *Chronicle of Higher Education* 48 (September 28, 2001): A39.
16. Cited in Ting Shi, "Dreams Deferred," *The New World*. http://journalism.berkeley.edu/ngo/reports/newworld/foreign.html (accessed January 1, 2007).
17. Jenna Russell, "Foreign Students Protest Fee at UMass, Part of Charge Meant to Fund Monitoring," *Boston Globe*, March 21, 2004. https://mail.uark.edu/msg_fs.html?&security=false7lang=en&popupLevel=undefined&ch...181 (accessed January 1, 2007).
18. *International Students, SEVIS by the Numbers*, March 31, 2006. http://www.ice.gov/sevis/ (accessed January 1, 2007).
19. Ibid.

20. David L. Wheeler, "Boom Times for Border Crossing," *Chronicle of Higher Education* 49 (November 22, 2002). http://chronicle.com/weekly/v49/i13/13a)6201.htm.
21. Stephen Burd, "Bush May Bar Foreign Students from 'Sensitive Courses,'" *Chronicle of Higher Education* 48 (April 26, 2002): A26.
22. Burd, "Bush May Bar Foreign Students," p. A26.
23. Ibid.
24. Ibid.
25. Victor Johnson, "The Perils of Homeland Security," *Chronicle of Higher Education* 49 (April 11, 2003): B7.
26. Ibid.
27. Ibid.
28. Daniel Walfish, "Kept Out, Foreign Students Find It a Difficult Year to Win the State Department's Favor," *Chronicle of Higher Education* 49 (November 15, 2002): A40.
29. Cited in Paul Mooney and Shailaja Neelakantan, "No Longer Dreaming of America," *Chronicle of Higher Education* 51 (October 8, 2004): A41.
30. Quoted in Walfish, "Kept Out, Foreign Students," p. A40.
31. Quoted in Jennifer Jacobson, "Foreign-Student Enrollment Stagnates," *Chronicle of Higher Education* 50 (November 7, 2003): A1.
32. Mooney and Neelakantan, "No Longer Dreaming of America," p. A41.
33. Yudhijit Bhattacharjee, "Foreign Graduate Student Applications Drop," *Science* 303 (March 5, 2004): 1453.
34. Burton Bollag, "Wanted: Foreign Students," *Chronicle of Higher Education* 51 (October 8, 2004): A37.
35. John Gravois, "Admission of Foreign Students to American Graduate Schools Continues Its Post 9/11 Decline," *Chronicle of Higher Education* 51 (September 17, 2004): A32.
36. Burton Bollag, "Enrollment of Foreign Students Drops in U.S.," *Chronicle of Higher Education* 51 (November 19, 2004): A1.
37. Council of Graduate Schools, *Council of Graduate Schools Finds Decline in New International Graduate Student Enrollment for the Third Consecutive Year* (Washington, DC: The Council, 2004), p. 1. See also, Ed Frauenheim, "Foreign Student Enrollment Declines," CNET News, November 10, 2004, http://news.com/Foreign-student-enrollment+declines/2100-1088_3-5447691.html (accessed January 1, 2007).
38. Consult http://www.cgsnet.org/pdf/Sept04/FinalIntAsmissionsSurveyReport.pdf. NAFSA: Association of International Educators et al., *Survey of Foreign Student and Scholar Enrollment and Visa Trends for Fall 2004* (November 2004), p. 1.
39. NAFSA, *Survey of Foreign Student*, p. 2.
40. Ibid.
41. Heath Brown and Maria Doulis, *Findings from 2005 CGS International Graduate Admissions Survey I* (Washington, DC: Council of Graduate Schools, 2005).
42. Quoted in Brown and Doulis, *Findings from 2005 CGS*, p. 2.

43. NAFSA, *Survey of Foreign Student*, p. 2.
44. Ibid.
45. The full text of Altbach's remarks are supplied at http://www.bc.edu/bc_org/avp/soe/cihe.
46. Quoted in NAFSA, *Survey of Foreign Student*, p. 3.
47. "You Can't Get Here From There," *New York Times*, November 29, 2004, 29.
48. Bonnie Bibbs Kumm, "Coming to America? Think Again," *Community College Week* 15 (August 19, 2002): 5. See also, S. S. Barger, *The Impact of International Students on Students in U.S. Institutions of Higher Education* (Madison, WI: University of Wisconsin–Madison, PhD diss., 2004).
49. J. Michael Adams, "The Cost of Closing Our Doors," *North Jersey Herald*, November 28, 2004. http"//opendoors.iienetwork.org/?p=53075 (accessed January 1, 2007).
50. Johnson, "The Perils of Homeland Security," p. B7.
51. Fareed Zakaria, "Rejecting the Next Bill Gates," *Newsweek*, November 28, 2004. http://msnbc.msn.com/id/6542347/site/newsweek/ (accessed January 1, 2007).
52. "Keep the Brains Coming In," *Boston Herald*, February 28, 2005; and "Editorial: Foreign Student, Keep Them Coming to America," *Philadelphia Inquirer*, December 27, 2004, http://opendoors.iienetwork.org/?p=53075 (accessed January 1, 2007).
53. "Fewer Foreign Students, Less Security at Home," *Daytona Beach News Journal*, November 26, 2004, 13.
54. "Security Rules Hurt Colleges' Ability to Attract Bright Foreign Students," *Indianapolis Star*, December 5, 2004. http://www.indystar.com/articles/7/19595-1477-021.html (accessed January 1, 2007).
55. "We Need Foreign Students," *Cincinnati Post*, November 30, 2004. http://www.cincypost.com/2004/11/30/editbl113004.html.
56. "You Can't Get Here From There," *New York Times*, p. 29.
57. "Drop in Numbers Represents Lost Opportunity," *Detroit Free Press*, November 23, 2004. http://opendoors.iienetwork.org?p=53075 (accessed January 1, 2007).
58. Quoted in Tony Best, "Closing Doors of U.S. Colleges and Universities to Foreign Students," *CaribNews*, November 16, 2004, 1.
59. Beth McMurtrie, "Recruiting the World, American Universities Step Up Their Sales Pitch," *Chronicle of Higher Education* 51 (February 11, 2005). http://chronicle.com/weekly/v51/i23/23a00801.html.
60. Jamilah Evelyn, "Community Colleges Go Globe-Trotting," *Chronicle of Higher Education* 51 (February 11, 2008).
61. Lloyd Armstrong Jr., "Globalization and Higher Education," *Navigator* 6 (Fall 2006): 3.
62. Robert Sedgwick, "The Bologna Process: As Seen from the Outside," *World Education News and Reviews* (September/October 2003): 1.
63. Ibid.
64. Philip G. Altbach, "Education Mecca," *Connection: New England's Journal of Higher Education* (Spring 2005): 1–2. https://mail.uark.edu/msg_fs.htmil (accessed January 1, 2007).

65. Hey-Kyung Koh Chin, "The New Landscape of International Student Mobility," *International Higher Education* 43 (Spring 2006): 9. Refer also to Burton Bollag, "Enrollment of Foreign Students Holds Steady," *Chronicle of Higher Education* (November 17, 2006). http://chronicle.com/temp/email2.php?id=pnzb5TrvMbTykWsyzzVsCWMPwwH (accessed January 1, 2007).
66. Eugene McCormack, "Foreign Graduate Students Flock to U.S. in Greater Numbers, Report Says," *Chronicle of Higher Education* 53 (September 1, 2006): A68.
67. David Cohen, "Foreign Students' Interest in the United States Shrinks, Study Finds," *Chronicle of Higher Education* 53 (October 27, 2006): A41.
68. Quoted in Bollag, "Enrollment of Foreign Students," p. A37.
69. Burton Bollag and Kelly Field, "Foreign Students: Uncle Sam Wants You," *Chronicle of Higher Education* 52 (January 20, 2006): A45.
70. Ibid.
71. *Under Secretary of State for Public Diplomacy and Public Affairs Karen Hughes at the Summit of U.S. Presidents on Higher Education.* http://whitehouse.gov/news/releases/2006/01 (accessed January 1, 2007).

Epilogue

1. See George Borjas, *An Evaluation of the Foreign Student Program*, Center for Immigration Studies, June 2002. http://www/cis.org/articles/2002/back602.html (accessed January 1, 2007).
2. George Borjas, "Rethinking Foreign Students," *National Review*, June 2, 2002. http://www.nationalreview.com/issue/borjas061702.asp (accessed January 1, 2007).
3. The discussion transcript was circulated by the Center for Immigration Studies and posted at http://www.cis.org/articles/2002/back602.html. Refer to Catherine E. Shoichet, "Student Visa Policy Forum: Panelists Clash Over Value of Foreign Students at American Colleges," *Chronicle of Higher Education* (June 26, 2002). http://opendoors.iienetwork.org/?p=29520 (accessed January 1, 2007).
4. All subsequent Borjas references and direct quotes, except as noted, are reproduced from Borjas, "Rethinking Foreign Students."
5. The argument, of course, was advanced prior to the full development of the SEVIS database.
6. Writing for the *Boston Globe* (September 15, 2002), Borjas reiterated part of his argument as follows: "No one seems to be keeping track of what these students are studying. We ban the export of goods considered vital to our national security, such as materials used to produce weapons of mass destruction. Yet there is no similar ban on the type of knowledge that can be acquired in American universities and exported abroad." http://www.ksg.harvard.edu/news/opeds/2002/borgas_student_ program_bg_091502.htm (accessed January 1, 2007).

7. Reported in the IIE Network. "Panelists Clash Over Value of Foreign Students at American Colleges," *Chronicle of Higher Education* (June 26, 2002): 1. http://opendoors.iienetwork.org/?p=29520 (accessed January 1, 2007).
8. Terry W. Hartle, "Symposium-Evaluating Subsidies to Foreign Students in the U.S.," *Insight on the News* (August 26, 2002). http://www.findarticles.com/p/articles/mi_m1571/is_31_18_ai90990437 (accessed January 1, 2007).
9. Stephen Dinan, *Washington Times*, June 26, 2002. http:/opendoors.iienet-work.org/?p=29519 (accessed January 1, 2007).
10. Allen E. Goodman, "In Response to 'Questioning the Foreign Student Program,'" *Washington Post*, June 26, 2002, 1.
11. Ibid.
12. George Borjas, "Do Foreign Students Crowd Out Native Students from Graduate Programs?" National Bureau of Economic Research Working Paper No. W10349 (March 2004). Social Science Research Network. http://ssm.com/abstract=515243 (accessed January 1, 2007).
13. George Borjas, "Immigration in High-Skill Labor Markets: The Impact of Foreign Students on the Earnings of Doctorates," National Bureau of Economic Research Working Paper No.12085 (March 2006). Abstract, http://www.nber.org/papers/w12085.
14. George Borjas, *The Labor Market Impact of High-Skill Immigration* (December 2004), manuscript available from the author at gborjas@harvard.edu.
15. Cited in part by Jeanne Batalova, "The 'Brain Gain' Race Begins with Foreign Students," *Migration Policy Institute* (January 1, 2007). Migration Information Service, source@migration.information.org; also available as a working paper (dated September 2006) from the Immigration Policy Center.
16. See *Open Doors 2006, Report on International Education Exchange* (New York: Institute of International Education, 2006).
17. Stuart Anderson, "International Students and U.S. Policy Choices," *International Educator* (November–December 2005): 25.
18. Marlene M. Johnson, "Toward a New Foreign-Student Strategy," *Chronicle Review* 52 (July 25, 2006): B16.
19. NAFSA: Association of International Educators, *Restoring U.S. Competitiveness for International Students and Scholars* (June 2006): 1; also refer to the January 2003 NAFSA Report, *In America's Interest: Welcoming International Students*.
20. NAFSA, *Restoring U.S. Competitiveness*, pp. 1–2.
21. Ibid., p. 4.
22. Johnson, *Toward a New Foreign-Student Strategy*, p. B16.
23. NAFSA, *Restoring U.S. Competitiveness*, p. 5.
24. Johnson, *Toward a New Foreign-Student Strategy*, p. B16.
25. Batalova, "The 'Brain Gain' Race," p. 4.
26. Ibid., p. 7.
27. NAFSA, *Restoring U.S. Competitiveness*, p. 10.
28. Andrea G. Trice, "Faculty Perceptions of Graduate International Students: The Benefits and Challenges," *Journal of Studies in International Education* 7 (Winter 2003): 379–403.

29. Andrea G. Trice, "Navigating in a Multinational Learning Community: Academic Departments' Responses to Graduate International Students," *Journal of Studies in International Education* 9 (Spring 2005): 62–89.
30. Roger Yao Klomegah, "Social Factors Relating to Alienation Experienced by International Students in the United States," *College Student Journal* 40 (June 2006): 303–15.
31. Ibid., p. 313.
32. Shideh Hanassab, "Diversity, International Students, and Perceived Discrimination: Implications for Educators and Counselors," *Journal of Studies in International Education* 10 (Summer 2006): 157–172.
33. Sevan G. Terzian and Leigh Ann Osborne, "Postwar Era Precedent and the Ambivalent Quest for International Students at the University of Florida," *Journal of Studies in International Education* 10 (Fall 2006): 300–301.
34. NAFSA, *Restoring U.S. Competitiveness*, p. 3.

Bibliography

Adams, J. Michael. "The Cost of Closing Our Doors." *North Jersey Herald*, November 28, 2004. http://opendoors.iienetwork.org/?p=53075.
Allen, Analee. "International House Marks Seventieth Year." University of California at Berkeley. http://ias.berkeley.edu/ihouse/I/tribune.html (accessed January 1, 2007).
Altbach, Philip G. "Higher Education Crosses Borders." *Change* 36, no. 2 (March–April, 2004): 18–19.
Altbach, Philip, David Kelly, and Y. Lulat. *Research on Foreign Students and International Study*. New York: Praeger, 1985.
Altbach, Philip, and Jing Wang. *Foreign Students and International Study 1984–1988*. Lanham, MD: University Press of America, 1989.
Altbach, Philip G. "Education Mecca." *Connection: New England's Journal of Higher Education* (Spring 2005): 1–2. https://mail.uark.edu/msg_fs.htmil.
Althen, Gary. *The Handbook of Foreign Student Advising*. Washington, DC: NAFSA, 1984.
American Council on International Intercultural Education and the Stanley Foundation. "Building the Global Community: The Next Step." November 28–30, 1994. http://www.stanleyfoundation.org/reports/CC1.pdf.
AMIDEAST. *1951–2006: 50 Years of Bridging Cultures and Building Understanding*. Washington, DC. http://www.amideast.org/about/50_years/default.htm (accessed January 1, 2007).
Anbarasan, Ethiraian. "India's loss, West's gain — Higher Education . . . and After?" *UNESCO Courier* (September 1998).
Anderson, Stuart. "International Students and U.S. Policy Choices." *International Educator* (November–December 2005): 25.
Ariff, J. Oral. "Interview with A. C. Blaisdell (1928–1961)." In *Foreign Students and Berkeley International House*. University of California at Berkeley Press, 1967.
Armstrong, Lloyd Jr. "Globalization and Higher Education." *The Navigator* 6 (Fall 2006): 3.
Arndt, Richard T., and David Lee Rubin, eds. *The Fulbright Difference 1948–1992: Studies on Cultural Diplomacy and the Fulbright Experience*. New Brunswick, NJ: Transaction Publishers, 1993.
Bain, Olga, and William K. Cummings. "Where Have the International Students Gone?" *International Instructor* 14 (March–April 2005).

Baker, Billy W. "The Foreign Transfer Student: Feel and Fact." *College and University* 50 (Spring 1975).

Bananal, Eduardo. *Camilo Osias: Educator and Statesman*. Quezon City, PI: Manlapaz Publishing Co., 1974.

Batalova, Jeanne. "The 'Brain Gain' Race begins with Foreign Students." *Migration Policy Institute*. Available online (accessed January 1, 2007) from the Migration Information Service at source@migration.information.org and as a working paper dated September 2006 from the Immigration Policy Center.

Bera, Anil K. "Cosmopolitan Club, Tagores, and UIUC: A Brief History of 100 Years." *Cosmo Connections*, Spring 2006, 100th anniversary ed. http://netfiles.uiuc.edu/ro/www/CosmopolitanClubattheUniversityofIllinois/connections/nlspr06/nlspr06a.html (accessed January 1, 2007).

Berthoff, Warner. "George Ticknor: Brief Life of a Scholarly Pioneer 1791–1871." *Harvard Magazine* (January–February, 2005).

Bevis, Teresa Brawner. *Attitudes and Characteristics of Foreign Freshmen at the University of Arkansas*. Unpub. PhD diss., 2001, pp. 62–78.

Bhattacharjee, Yudhijit. "Foreign Graduate Student Applications Drop." *Science* 303 (March 5, 2004): 1453.

Biggs, James. *History of Don Francisco de Miranda's Attempt to Effect a Revolution in South America*. Biography by W. S. Robertson. London: printed for the author by T. Gillet, 1809.

Bogdanski, John A. "The IRS Imposes a Ridiculous Burden on Foreign Students." *Chronicle of Higher Education* (May 2, 1997).

Boggs, George R. "About AACC." American Association of Community Colleges, 2006. http://www.aacc.nche.edu.

Bollag, Burton. "Enrollment of Foreign Students Drops in U.S." *Chronicle of Higher Education* 51 (November 19, 2004).

Bollag, Burton. "Enrollment of Foreign Students Holds Steady." *Chronicle of Higher Education* (November 17, 2006).

Bollag, Burton. "Wanted: Foreign Students." *Chronicle of Higher Education* 51 (October 8, 2004).

Bollag, Burton, and Kelly Field. "Foreign Students: Uncle Sam Wants You." *Chronicle of Higher Education* 52 (January 20, 2006).

Borjas, George. "Immigration in High-Skill Labor Markets: The Impact of Foreign Students on the Earnings of Doctorates." March 2006. National Bureau of Economic Research Working Paper No.12085.

Borjas, George. *The Labor Market Impact of High-Skill Immigration*. December 2004. Manuscript available from the author at gborjas@harvard.edu.

Borjas, George. *An Evaluation of the Foreign Student Program*. Center for Immigration Studies, June 2002. http://www/cis.org/articles/2002/back602.html.

Borjas, George. "Rethinking Foreign Students." *National Review* (June 2, 2002). http://www.nationalreview.com/issue/borjas061702.asp.

Borjas, George. "Do Foreign Students Crowd Out Native Students from Graduate Programs?" March 2004. National Bureau of Economic Research Working Paper No. W10349. Available from Social Science Research Network. http://ssm.com/abstract=515243.

Braziel, Jana. "History of Migration and Immigration Laws in the United States." University of Massachusetts. http://www.umass.edu/complit/aclanet/USMigrat.html (accessed January 1, 2007).

Breuder, Robert L. *A Statewide Study: Identified Problems of International Students Enrolled in Public Community/Junior Colleges in Florida*. Florida State University and the University of Florida: Center for State and Regional Leadership, May 1972. Eric Microfiche Collection, No. ED 062 977.

Brickman, William M. "Historical Development of Governmental Interest in International Higher Education." In *Governmental Policy and International Education*, edited by Stewart Fraser. New York: John Wiley & Sons, Inc., 1965.

Brown, H., and M. Doulis. *Findings from 2005 CGS International Graduate Admissions Survey I*. Washington, DC: Council of Graduate Schools, 2005.

Burd, Stephen, "Bush May Bar Foreign Students from 'Sensitive Courses.'" *Chronicle of Higher Education* 48 (April 26, 2002): A26.

Caldwell, Oliver J. "Education Comes of Age Around the World." In *Governmental Policy and International Education*, edited by Stewart Frazer. New York: John Wiley & Sons, 1965.

Capes, W. W. *University Life in Ancient Athens*. New York: G. E. Stechert and Company, 1922.

Chaing Yung-chen. *Chinese Studies in History* 36, no. 3 (Spring 2003).

Chalmers, Paul M. "The Founding and Growth of NAFSA." *Institute of International Education News Bulletin* XXVI, 7 (April 1, 1951): 32.

Chase, Audree, and James R. Mahoney eds. *Global Awareness in Community Colleges, A Report on a National Survey*. Washington, DC: American Association of Community Colleges, 1996.

Cieslak, Edward Charnwood. *The Foreign Student in American Colleges*. Detroit: Wayne University Press, 1955.

Clarke, Helen I., and Martha Ozawa. *The Foreign Student in the United States*. Madison, WI: School of Social Work, University of Wisconsin, 1970.

Clinton, Bill. *My Life*. New York: Alfred A. Knopf, 2004, p. 98.

Cohen, David. "Foreign Students' Interest in the United States Shrinks, Study Finds." *Chronicle of Higher Education* 53 (October 27, 2006).

College Entrance Examination Board. "The Foreign Student in United States Community and Junior Colleges"; A Colloquium Held at Wingspread, Racine, Wisconsin. October 18–20, 1977. ERIC: ED154858.

Columbia University. "Mario Garcia Menocal." *The Columbia Encyclopedia*. 6th ed. New York: Columbia University Press, 2001.

Committee on Educational Interchange Policy. *Expanding University Enrollments and the Foreign Student*. New York: Author, 1957.

Committee on Educational Interchange Policy. *Chinese Students in the United States, 1948–1955*. New York: Author, 1956.

Committee on Friendly Relations Among Foreign Students. *Unofficial Ambassadors*. New York: Author, 1945.

Compayre, Gabriel. *Abelard and the Origin and Early History of Universities*. New York: Charles Scribner's Sons, 1910.

Coulter, Ann. "Are You Now or Have You Ever Been a Second-Rate Filmmaker?" Townhall.com (November 17, 2005).

Council of Graduate Schools. *Council of Graduate Schools Finds Decline in New International Graduate Student Enrollment for the Third Consecutive Year.* Washington, DC: Council of Graduate Schools, 2004, p. 1.

Courtney, Steve. "Rising Power in the East: Joe Twichell, Mark Twain and the Chinese Educational Mission, 1872–81." Unpublished research. Hartford, CY: Asylum Hill Congregational Church archives. http://www.ahcc.org/TWAIN/MarkJoeandBOYS4.htm (accessed January 1, 2007).

Cowan, Geoffrey. "Hughes Offers Steps, Not Spin." *USA Today*, September 29, 2005: 12A.

Da Ponte, Durant. The Fulbright Program in Spain. *Hispania*, 44, no. 3 (September 1961).

Daly, Lloyd W. "Roman Study Abroad." *American Journal of Philology* LXXI: 1950.

Davis, Jerome Dean. *A Sketch of the Life of Reverend Joseph Hardy Neesima*. Kyoto, Japan: Doshiva University, 1936.

Desruisseaux, Paul. "2-Year Colleges at Crest of Wave in U.S. Enrollment by Foreign Students." *Chronicle of Higher Education* 45, no. 16 (December 11, 1998).

Desruisseaux, Paul. "Intense Competition for Foreign Students Sparks Concerns About U.S. Standing." *Chronicle of Higher Education* 45, no. 7 (October 9, 1998).

Douglass, Sara. *Pilgrims, Students, and Gentlemen*. 2000 http://www.saradouglass.com/1570.html (accessed January 1, 2007).

Du Bois, Cora. *Foreign Students and Higher Education in the United States*. Washington, DC: American Council on Education, 1956. http://chronicle.com/weekly/v51/i23/23a01101.htm.

Engel, Julien. "The African-American Institute." *African Studies Bulletin* 6, no. 3 (October 1963).

Evangelista, Susan. *Carlos Bulosan and His Poetry: A Biography and Anthology*. Seattle: University of Washington Press, 1985.

Fidler, Isaac. *Observations on Professions, Literature, Manners and Emigration in the United States and Canada Made During a Residence There in 1832*. London: Whittaker, Treacher, and Company, 1833.

Fraser, Stewart, and William W. Brickman. *A History of International and Comparative Education* (Chicago: Scott Foresman & Company, 1968), p. 294.

Fraser, Stewart. *Government Policy and International Education*. New York: Wiley & Sons, Inc., 1962.

Frauenheim, Ed. "Foreign Student Enrollment Declines." CNET News, November 10, 2004. http://news.com/Foreign-student-enrollment+declines/2100-1088_3-5447691.html.

Fryer, John. "Admission of Chinese Students to American Colleges." *United States Bureau of Education*, Bulletin no. 2, whole no. 399. Washington DC: Government Printing Office, 1909.

Glazier, L., and L. Kenschaft. "Welcome to America." *International Educator*, XI, no. 3 (Summer 2002).

Goff, Janet E. "Tribute to a Teacher: Uchimura Kanzo's Letter to William Smith Clark." *Monumenta Nipponica* 43, no. 1 (Spring 1988).

Goodman, Allen E. "In Response to 'Questioning the Foreign Student Program.'" *Washington Post*, June 26, 2002: 1.

Goodman, Allen E. "What Foreign Students Contribute." *Chronicle of Higher Education* 42, no. 23 (February 16, 1996).

Golden, Daniel. "Foreign Students' High Tuition Spurs Junior Colleges to Fudge Facts, The Boston College Center for International Higher Education." *Higher Education* (Fall 2002).

Gravois, John. "Admission of Foreign Students to American Graduate Schools Continues Its Post 9/11 Decline." *The Chronicle of Higher Education* 51 (September 17, 2004).

Hanassab, Shideh. "Diversity, International Students, and Perceived Discrimination: Implications for Educators and Counselors." *Journal of Studies in International Education* 10 (Summer 2006).

Haskins, Charles Homer. *The Renaissance of the Twelfth Century*. Cambridge, MA: Harvard University Press, 1928.

Hebel, Sara. "Proposed Rules on Foreign Students Leave Many Colleges Worried." *Chronicle of Higher Education* (October 26, 2001): A29.

Heller, Nathan J., and Jessica R., Rubin-Wills (*Crimson* Staff Writers). "Trying Times, Harvard Takes Safe Road." *Harvard Crimson*, June 5, 2003.

Hey-Kyung Koh Chin, "The New Landscape of International Student Mobility." *International Higher Education* 43 (Spring 2006): 9.

Hibbert, Christopher. *The Grand Tour*. New York: G. P. Putnam's Sons, 1969.

Hooker, Richard. *Ch'ing China: The Boxer Rebellion*. Washington State University, 1996.

Houser, George M. "Meeting Africa's Challenge: The Story of the American Committee on Africa." *Issue: A Journal of Opinion* 6, no. 2–3 (Summer-Autumn 1976): 16–26.

Huang, Hui. "Overseas Studies and the Rise of Foreign Cultural Capital in Modern China." *International Sociology* 17, no. 1 (2002).

Hung, William. "Huang Tsun-Hsien's Poem 'The Closure of the Educational Mission in America.'" *Harvard Journal of Asiatic Studies* 18, no. 1–2 (June 1955).

Institute of International Education. *Blueprint for Understanding*. New York: Author, 1949.

Institute of International Education. *Open Doors 1955–1956*. Report on International Educational Exchange. New York: Author.

Institute of International Education. *Open Doors, 1948–2006*. Report on International Education Exchange. New York: Author.

Jacobson, Jennifer. "Foreign-Student Enrollment Stagnates." *Chronicle of Higher Education* 50 (November 7, 2003): A1.

Jamilah, Evelyn. "Community Colleges Go Globe-Trotting." *Chronicle of Higher Education* 51 (February 11, 2008).

Jefferson, Thomas. *Writings: Autobiography, Notes on the State of Virginia, Public and Private Papers, Addresses, Letters*. ed. Merrill D. Peterson. New York: Library of America, 1994.

Johnson, Jean M., and Mark Regets. *International Mobility of Scientists and Engineers to the United States: Brain Drain or Brain Circulation?*. Arlington, VA: National Science Foundation (NSF 98-316), 1998.

Johnson, Marlene M. "Toward a New Foreign-Student Strategy." *Chronicle Review* 52 (July 25, 2006).

Johnson, Marlene M. "Transforming the National Consciousness." *International Educator* 14, no. 5 (September–October 2005): 37.

Johnson, Victor C. "The Perils of Homeland Security." *Chronicle of Higher Education* 49 (April 11, 2003).

Johnson, Victor C. "Taking Stock, Making Strides." *International Educator* 14, no. 1 (March–April 2005).

Kandel, Isaac. "United States Activities in International Cultural Relations." *American Council on Education Studies*, series I, vol. IX, no. 23. Washington: American Council on Education, 1945.

Karnow, Stanley. *In Our Image: America's Empire in the Philippines.* New York: Random House, 1990.

Kibre, Pearl. "Nations in the Mediaeval Universities." *Mediaeval Academy of America* (1948): 49.

Kronenwetter, Eric. "U.S. Remains Unpopular Despite Efforts to Repair Image Abroad," *International Educator* 14, no. 4 (July–August 2005): 8.

King, Maxwell C., and Robert L. Breuder. *New Directions for Community Colleges*, editor's notes, 1979, 26.

Klomegah, Roger Yao. "Social factors Relating to Alienation Experienced by International Students in the United States." *College Student Journal* 40 (June 2006).

Koh, Harold Hongju. "Yellow in a White World." *Yale Global* (September 27, 2004). Yale Center for the Study of Globalization. http://yaleglobal.yale.edu/display.article?id=4653 (accessed January 30, 2007).

Koos, Leonard V. "Recent Growth of the Junior College" (1928). Reprinted in *American Journal of Education* 91, no. 4, *The First 90 Years* (August 1983).

Kumm, Bonnie Bibbs. "Coming to America? Think Again." *Community College Week* 15 (August 19, 2002).

Liping Bu. "The Challenge of Race Relations: American Ecumenism and Foreign Student Nationalism, 1900–1940." *Journal of American Studies* 35 (2001). Cambridge University Press.

Lochner, Louis B. "Cosmopolitan Clubs in American University Life/" *American Review of Reviews* 37 (1908).

Lucas, Christopher J. *American Higher Education: A History*, 2nd ed. New York: Palgrave Macmillan, 2006.

Marion, Paul B. "Research on Foreign Students at Colleges and Universities in the United States." *New Directions for Student Services* 1986, no. 36.

Marrou, H. I. *A History of Education in Antiquity*. New York: Mentor, 1964.

Marshall, Thomas. "The Strategy of International Exchange." In *Global Opinion, The Spread of Anti-Americanism*. Washington DC: Pew Research Center, January 24, 2005.

Martorana, S. V. "Constraints and Issues in Planning and Implementing Programs for Foreign Students in Community and Junior Colleges." In *The Foreign Student in United States Community and Junior Colleges.* New York: College Entrance Examination Board, 1978.

Maurer, Isabel A. "The Fulbright Act in Operation." *Far Eastern Survey* 18, 9 (May 4, 1949).
McCormack, Eugene. "Foreign Graduate Students Flock to U.S. in Greater Numbers, Report Says." *Chronicle of Higher Education* 53 (September 1, 2006).
McMurtrie, Beth. "Recruiting the World, American Universities Step Up Their Sales Pitch." *Chronicle of Higher Education* 51 (February 11, 2005).
McMurtrie, Beth. "American Colleges Experience a Surge in Enrollments from India." *Chronicle of Higher Education* 48, no. 12 (December 10, 1999).
McMurtrie, Beth. "Commission Seeks to Prevent Terrorism by Monitoring Foreign Students." *Chronicle of Higher Education* 46 (June 16, 2000).
Mead, William E. *The Grand Tour in the Eighteenth Century.* New York: Benjamin Bloom, 1972.
Meloni, Christine F. "Adjustment Problems of Foreign Students in U.S. Colleges and Universities." ERIC #: ED276296.
Mooney, Paul, and Shailaja Neelakantan. "No Longer Dreaming of America." *Chronicle of Higher Education* 51 (October 8, 2004).
Murphy, Christopher. "Going with the Flows?" *International Educator* 14 (March–April 2005): 2.
NAFSA: Association of International Educators. *Restoring U.S. Competitiveness for International Students and Scholars.* June 2006.
Neal, Joe W. "The Office of the Foreign Student Adviser." *Institute of International Education News Bulletin* XVII, no. 5 (February 1, 1952): 37.
Niefeld, S. J., and Harold Mendelsohn. "How Effective Is Our Student-Exchange Program?" *Educational Research Bulletin* 33, no. 2 (February 10, 1954).
Ong, Anna Liza R. *First Filipino Women Physicians.* Society of Philippine Health History, 2004. http://www.doh.gov.ph/sphh/filipino/women.htm (accessed January 1, 2007).
Pascoe, Peggy. "At America's Gates: Chinese Immigration during the Exclusion Era, 1882–1943 (review)." *Journal of Social History* 38, no. 3 (Spring 2005).
Pew Global Attitudes Project. *U.S. Image Up Slightly But Still Negative.* Washington DC: Pew Research Center, June 23, 2005.
Pitkin, Thomas M. *Keepers of the Gate: A History of Ellis Island.* New York: University Press, 1975.
Posadas, Barbara M. *The Filipino Americans.* The New Americans Series. Westport, CT: Greenwood Publishing Group, 1999.
Qian Ning, *Chinese Students Encountering America.* Translated by T. K. Chu. University of Washington Press, 2002.
Randall, John Herman. *The Making of the Modern Mind.* Cambridge, MA: Riverside Press, 1940.
Rhodes, E. J. M. "In the Shadow of Yung Wing." *Pacific Historical Review* 74 (February 2005).
Richards, Pamela Spence. "Cold War Librarianship: Soviet and American Library Activities in Support of National Foreign Policy, 1946–1991."*Libraries & Culture* 36, no. 1 (Winter 2001).
Rockefeller, David. *Memoirs.* New York: Random House, 2002.

Ross, Evan, "Friendships Traded in Gaps of War, Ideology International House History Before, After Wartime." *The Daily Californian* (Berkeley), 2005. dailycalifornian@dailycal.org.

Rowling, Marjorie. *Life in Medieval Times*. Berkley: Berkley Publishing Group, 1979.

Rubin, Amy Magaro. "U.S. Tests a New System to Track Foreign Students." *Chronicle of Higher Education* (September 19, 1997).

Rudolph, Frederick. *The American College and University*. Athens and London: The University of Georgia Press, 1990, p. 351.

Russell, Jenna. "Foreign Students Protest Fee at UMass, Part of Charge Meant to Fund Monitoring." *Boston Globe*, March 21, 2004.

Sadler, Michael E. "Impressions of American Education." *Educational Review* XXV (March 1903).

Schiff, Judith Ann. "When East Meets West." *Yale Alumni Magazine* (November/December 2004). Yale University Library.

Schrecker, Ellen. *The Age of McCarthyism*. New York: Bedford/St. Martin, 2005.

Sedgwick, Robert. "The Bologna Process: As Seen from the Outside." *World Education News and Reviews* (September/October 2003).

Sedgwick, Robert. "Middle Eastern Students Find Options at Home and Elsewhere." *World Education News and Reviews* (November/December 2004).

Selltiz, C., and S. W. Cook. "Factors Influencing Attitudes of Foreign Students Toward the Host Country." *Journal of Social Issues* 18 (1962).

Shoichet, Catherine E. "Student Visa Policy Forum, 'Panelists Clash Over Value of Foreign Students at American Colleges.'" *Chronicle of Higher Education* (June 26, 2002). http://opendoors.iienetwork.org/?p=29520.

Simon, Joan. *Education and Society in Tudor England*. London: Cambridge University Press, 1966.

Smalley, Martha Lund. *Guide to the Archives of the Chinese Students' Christian Association in North America*. New Haven, CT: Yale University Library, 1983.

Smith, M. Brewster. "Features of Foreign Student Adjustment." *Journal of Higher Education* 26, no. 5 (1955).

Smith, Marion. "Overview of INS History." In *A Historical Guide to the U.S. Government*, edited by George T. Kurian. New York: Oxford University Press, 1998.

Southwick, Ron. "College Officials Fear Tightening of Student-Visa Rules in Wake of Terrorist Attacks," *Chronicle of Higher Education* 48 (September 28, 2001).

Teng, Ssu-you, and J. K. Fairbank. *China's Response to the West: A Documentary Survey, 1839–1923*. Cambridge, MA: Harvard University Press, 1954.

Terzian, Sevan G., and Leigh Ann Osborne. "Postwar Era Precedent and the Ambivalent Quest for International Students at the University of Florida." *Journal of Studies in International Education* 10 (Fall 2006): 300–301.

Thorndike, Lynn. *University Records and Life in the Middle Ages*. New York: Columbia University Press, 1944.

Trice, Andrea G. "Faculty Perceptions of Graduate International Students: The Benefits and Challenges." *Journal of Studies in International Education* 7 (Winter 2003).

Trice, Andrea G. "Navigating in a Multinational Learning Community: Academic Departments' Responses to Graduate International Students." *Journal of Studies in International Education* 9 (Spring 2005).

U.S. Leadership in International Education. *The Lost Edge*. Conference Report and Action Agenda, United States Information Agency and Educational Testing Service. Washington, DC, 1998.

Waddell, Helen. *The Wandering Scholars*. London, 1927.

Walden, John W. H. *The Universities of Ancient Greece*. New York: Charles Scribner's Sons, 1909.

Walfish, Daniel. "Kept Out, Foreign Students Find it a Difficult Year to Win the State Department's Favor." *The Chronicle of Higher Education* 49 (November 15, 2002).

Wheeler, David L. "Boom Times for Border Crossing." *Chronicle of Higher Education* 49 (November 22, 2002).

Wheeler, W. Reginald, Henry H. King, and Alexander B. Davidson, eds. *The Foreign Student in America*. New York: Association Press, 1925.

White, Helen C. "An Integrated Cultural Relations Program." In *Special Report, Conference on International Student Exchanges, May 10, 11, 12, 1948, University of Michigan*, edited by Laurence Duggan. New York: Institute of International Education, September 22, 1948.

Wood, Charles T. *The Quest for Eternity, Medieval Manners and Morals*. New York: Doubleday Anchor, 1971.

Woodward, Colin. "At Community Colleges, Foreign Students Discover Affordable Degree Programs." *The Chronicle of Higher Education* 47, no. 12 (November 17, 2000).

Woolley, John, and Gerhard Peters. *The American Presidency Project*. Lyndon B. Johnson, Remarks to a Group of Foreign Students, May 5, 1964. Santa Barbara: University of California. http://www.presidency.ucsb.edu (accessed January 1, 2007).

Woolley, John, and Gerhard Peters. *The American Presidency Project*. Lyndon B. Johnson, Special Message to the Congress Proposing International Education and Health Programs, February 2, 1966. http://www.presidency.ucsb.edu (accessed January 1, 2007).

Woolley, John and Gerhard Peters. *The American Presidency Project*. John F. Kennedy, Remarks at a Reception for Foreign Students on the White House Lawn, May 10, 1962. Santa Barbara: University of California. http://www.presidency.ucsb.edu (accessed January 1, 2007).

Yung Wing. *My Life in China and America*. New York: Henry Holt & Co., 1909.

Zakaria, Fareed. "Rejecting the Next Bill Gates." *Newsweek*, November 28, 2004.

Zeszotarski, Paula. "Issues in Global Education Initiatives in the Community College." *Community College Review* 29 (Winter 2001).

Zhuxian, Yi. "Hu Shi, Chinese, and Western Culture." *The Journal of Asian Studies* 51, no. 4 (November 1992).

Zook, George. "The Residence of Foreign Students at American Universities." *United States Bureau of Education*. Bulletin (1915), no. 27, whole no. 654. Washington, DC.

Index

9/11 [September 11, 2001], 8, 11, 203, 206, 211, 219, 225

A
Abelard, Peter, 21
Acosta-Sison, Honoria, 76
African-American Institute, 161
African-American Students Foundation, 159
Altbach, Philip, 8, 185, 216, 223
Althen, Gary, 11
America-Mideast Educational and Training Services [AMIDEAST], 148, 150–151
American Association of Collegiate Registrars and Admissions Officers, 124, 207
American Association of Community Colleges [AACC], 180, 181
American Council on Education, xii, 8, 204, 206, 209
American Council on International Intercultural Education, 180, 181
Amherst College, 53–55
Andrus, J. Russell, 106, 107
anti-American sentiment, 4, 5
Antioch, 16, 17, 18
Aristotle, 15
Association of American Universities, 8, 209, 213, 216
Atoms-for-Peace, 136, 137

B
Barbour Scholarship, 90
Basic Naturalization Act [1906], 98
Benitez, Conrado F., 77, 78
Bolivar, Fernando, 41
Bolivar, Simon, 41
Bologna process, 222–223
Borjas, George, 231–236
Boxer Rebellion, 63, 94, 105, 126, 137
Boyer, Ernest, 179, 180
Brown, Ina Corinne, 1, 2, 4
Brown, Reverend Samuel Robbins, 43
Buisson, Ferdinand Eduard, 38
Bureau of Educational and Cultural Affairs, 108
Bureau of Immigration, 98
Bureau of Insular Affairs, 77
Burlingame Treaty, 45
Butler, Nicholas Murray, 95, 96

C
Caldwell, Oliver J., 1
Cappadocia, 17
Carnegie Endowment for International Peace, 101, 108, 110
Carnegie Foundation for the Advancement of Teaching, 179
Center for Immigration Studies, 231
Charles Stewart Mott Foundation, 164
Chartres, 21
Chaucer, 14
Chen Lan-pin, 46, 48, 49
China Foundation for the Promotion of Education and Culture, 65
China Institute in America, 127
Chinese Education Mission, 42, 48, 49, 51, 52, 53, 57
Chinese Exclusion Act [1882], 57

Christianity and foreign students, 87, 92
Cicero, 16
Cieslak, Edward C., xii, 22
Clarke, Helen C., xii, 1
Clemmons, Samuel [Mark Twain], 48, 50
Cold War, 131, 134, 149
colleges, antebellum, 31, 40
Columbia University, 69, 70, 78, 84, 92, 95, 96, 97, 117, 118, 122, 145, 152, 166, 192, 193
Commission on the Future of Community Colleges, 179
Committee on Educational Interchange Policy, 140
Committee on Friendly Relations Among Foreign Students, 6, 10, 60, 75, 79, 85, 88, 89, 90, 92, 143, 146
community colleges, 166–171
Community Colleges for International Development [CCID], 169
Conference of Allied Ministers of Education [CAME], 111
confessionalism, 26–27
Consortium for International Education, 169
Coombs, Philip H., 156, 158, 159, 161
Coordinated Interagency Partnership Regulating International Students [CIPRIS], 201, 202
Cosmopolitan Clubs, 10, 60, 75, 80–85, 88, 89, 90, 92
Council of Graduate Schools [CGS], 213, 216, 225
Council on International Educational Exchange [CIEE], 148, 149
Council for International Exchange of Scholars [CIES], 108
Council on Student Travel, 148, 149
credentials, evaluation of, 123–125
"crowd-out" effect, 235, 236

D
Davidson, Alexander B., xii
Displaced Persons Act [1948], 126

Doshisha University, 54
Du Bois, Cora, xii
Duggan, Laurence, 109–111, 122
Duggan, Stephen, 95–96, 101, 108, 109

E
Ellis Island, 56, 58
Emergency Committee in Aid of Displaced Foreign Scholars, 108
Emergency Committee in Aid of Displaced German Scholars, 108
English Language Training, xiii
English as a second language, 181
enrollments
  decline in, 161, 163–164, 180
  stabilization of, 166–167, 170–173
Epicurians, 15, 17
European Economic Area [EEA], 241
European Higher Education Area [EHEA], 222, 223

F
Fidler, Isaac, 37, 38
Filipino and Indian Naturalization Act [1946], 126
foreign students, attitudes and characteristics of, 185, 187, 188
foreign student adviser [FSA], 83, 88
foreign student exchange, best practices for, xiii
Foreign Student Summer Project, 112, 113
Franklin, Benjamin, 31, 32
Fulbright Educational Exchange Program, 103–108, 111, 132, 137, 148, 149
Fulbright-Hays Act, 104
Fulbright, J. William, 103, 104

G
George, Katy Boyd, 90
Germanic ideal, 31–35, 38
global competition, 175–177
Goliardic poetry, 20
Goodman, Allan, 177, 192

Göttingen, 31, 34, 35
Grand Tour, 28–29

H
Hamilton, Thomas, 36–37, 40
Harper, William Rainey, 167
Harvard University, 34–37, 47, 92, 109, 117, 118, 133, 145, 152, 161, 163, 183
Haskins, Charles, 19
Hausknecht, Emil, 38, 39
Hellenistic Greek education, 14, 15
homeland security, 208, 209
House Un-American Activities Committee, 110
Hubert H. Humphrey Fellowship Program, 162
Hunerwadel, Oscar and Helen, 107

I
Illegal Immigrant Reform and Immigrant Responsibility Act [IIRIRA] [1996], 201
Immigration Act [1882], 57
Immigration Act [1903], 98
Immigration and Nationality Act (INA) [1952], 133
Immigration and Naturalization Service [INS], 176, 196
immigration policies
  beginnings of, 55
  WWII era, 125, 126, 128
Institute of International Education [IIE], 6, 10, 11, 60, 93–97, 101, 102, 108–112, 117–122, 124
International Education Board, 70, 97
international exchange, rationale for, 231, 234, 244, 245
International Federation of Students, 82
International Graduate Admissions Survey, 215
International Houses, 84–88
isolationism, 60

J
Japan, first enrollments from, 53, 54, 55
Jefferson, Thomas, 32, 33, 34, 41
Johnson, Lyndon, 157, 158
Johnson, Marlene, 239

K
Kanzo, Uchimura, 54
Kassebaum, Nancy, Senator, 179
Kennedy, John F., 155, 156, 157, 159, 160
King, Henry H., xii
Klineberg, Otto, 3

L
Liang Tun-yen, 47
Linden Educational Services, 182, 183

M
Machiavelli, Niccolo, 26
Magnuson Act [1943], 126
McCarran-Walter Act [1952], 133, 134
McCarthy, Joseph, 131, 132
McCarthyism, 108, 110, 111, 133, 140
Menocal, Mario Garcia, 42
Michigan International Student Problem Inventory, 168
migration, intellectual, 189, 190, 191
Miranda, Francisco de, 40, 41, 62
Mott, Dr. John R., 88
Mundt, Karl E., 105, 110
Murrow, Edward R., 108, 109, 110, 111
museums, 16, 17
Mutual Educational and Cultural Exchange Act, 156

N
NAFSA: Association of International Educators, 6–9, 12, 122, 123, 148, 150, 202, 206, 210, 213, 216, 217
National Academy of Sciences, 8
National Association of State Universities and Land Grant Colleges, 209, 213

National Liaison Committee on Foreign Student Admissions, 168
Neesima, Joseph Hardy [Niijima Jo], 53, 54
Nixon, Richard M., 159, 160, 162
Northrop, Birdsey, 46
Nu, U, 106

O
Office of the Coordinator of Inter-American Affairs, 121
Oklahoma City bombing, 203
Organization of the Petroleum Exporting Countries [OPEC], 172, 173

P
pensionado program, 75–78
Philippine Columbian Association, 78
Philippine Organic Act [1902], 74
physicians, training of, 146–147
Plato, 15
Platonic Academy, 15
Point Four [program], 137
progressivism, 59
Protagoras, 14

Q
Qing [Ch'ing] dynasty, 45, 47

R
Refugee Relief Act [1953], 140
Revolutionary War, impact of, 33
rhetorical schools, 14, 16
Rockefeller Foundation, 108, 121
Rockefeller, John D. Jr., 70, 72, 84, 85–88, 96, 97
Roman education, 16–17
Root, Elihu, 95, 96
Rudolph, Frederick, 58, 59

S
Sadler, Michael E., 39, 40
Scholar Rescue Fund, 108
Schurman, Jacob Gould, 82
Seelye, Julius H., 53

segregation, 34
Shriver, [Robert] Sargent, 159
"slave education," 71, 73
Smith-Mundt Act, 105
Socrates, 14, 15
Sophist [Gr.: *sophisai* (pl.)], 14
South Africa Education Program [SAEP], 163
Stanley Foundation, 180–182
Strategic Taskforce on International Student Access, 8
student acculturation and adjustment, 141–143, 150, 171–172
student enrollments
  Chinese, 65, 66
  Filippino, 61, 74–78, 92
  Indian, 189, 194, 195
  Japanese, 79
  Latin American, 61–63
Student and Exchange Visitor Information System [SEVIS], 202, 206–208
Student and Exchange Visitor Program [SEVP], 208
students
  female, 76, 89, 90, 91
  misconduct of, 17, 18, 20, 25
Summit on U.S. Leadership in International Education, 176
Symonds, J.A., 19

T
Tagore, Rabindranath, 79
Tagore, Rathindranath, 79, 80
terrorism, 201–203, 211, 217, 220
Ticknor, George, 34–36
Timon, 17
Tsing Hua University, 65, 68, 69, 72
Tutu, Desmond, Archbishop, 163
Twichell, Joseph Hopkins, 48, 49, 50

U
Ugly American, 4, 107
United Nations, 103, 104, 110, 117

United Nations Educational, Scientific and Cultural Organization [UNESCO], 111, 112, 194
United States, criticism of, 4
United States Agency of International Development [USAID], 163
United States Bureau of Education (Office of Education), 61, 78, 93, 94, 114, 121, 124, 127
United States Department of State Bureau of Educational and Cultural Affairs, 162
United States Department of State Division of the American Republics, 109
United States Information and Educational Exchange Act, 105
United States State Department's Office of East-West Contacts, 148
universities, medieval, 20–27
USA Patriot Act, 206
USIA [United States Information Agency], 175, 176

V
visas
 issuance of, 231–233, 235–236, 239–240, 242
 restrictions on, 204, 206, 207, 211–213, 215–216, 218–220, 225, 228

W
Waldon, J.W.H., 15
Washington, George, 32–34, 36, 40
Wheeler, W. Reginald, xii
Wilson, Howard E., 6
World Student Christian Federation, 60
World Trade Center bombing, 195, 201–204
World University Science, 142

XYZ
Yale University, 41–44, 47, 48, 50–52, 62, 66, 67, 69, 75, 76, 92
Young Men's Christian Association [YMCA], 6, 80, 84, 89, 92, 93, 120
Young Women's Christian Association [YWCA], 6
Yung Wing, 42–52, 57
Zaroubin, Georgi N., 142